Sawmill Techniques for Southeast Asia

PROCEEDINGS OF THE FIRST
SOUTHEAST ASIA SAWMILL SEMINAR
SINGAPORE, JUNE 1975

Edited by Duane C. Mason
Associate Editor, World Wood

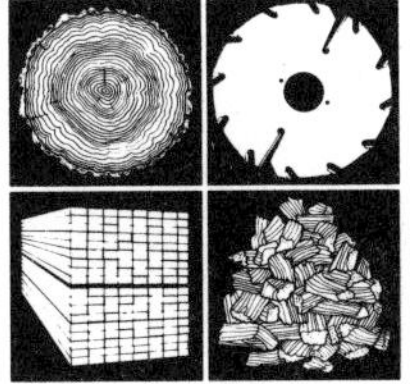

Miller Freeman Publications, Inc.

Contents

Contents

The forest resources of Southeast Asia are invaluable.

As might logically be assumed, this value is of a dual nature—serving domestic needs and providing an almost unparalleled commodity for world trade. In fact, there is every reason to expect that the latter exceeds the former in worth to the countries which possess these resources since returns from marketing them—whether in raw, semi-finished, or finished form—can only strengthen domestic vitality.

This volume contains the proceedings of a seminar whose goal was to show how to derive optimum benefits from these exceedingly valuable resources now and in the future.

This was the first Southeast Asia Sawmill Seminar, one in a series of seminars conducted by Miller Freeman Publications, San Francisco, California, USA, and sponsored by various of the company's forest-based business magazines.

The Southeast Asia Sawmill Seminar was sponsored by *World Wood* magazine and held at Singapore in June 1975. The Seminar's audience and many of its program participants represented not only that region of the world, but also Europe and North America. In all, more than 225 persons from many countries attended this event.

The thrust of the Seminar is well summarized in the title of Chapter 1: "Matching Southeast Asia's Wood Resources to Its Opportunities." That was the title of the opening presentation by H. C. "Carl" Mason, president of H. C. Mason & Associates, Inc., a consulting firm based at Gladstone, Oregon, USA. The firm's work has involved forest industry projects in nearly every major forest region of the world.

Mason is headquartered in the heart of what generally has been accepted as the most developed region in the world in processing large timber—the Pacific Northwest of North America (as used here, this embraces the province of British Columbia as well as the northwestern corner of the United States). Practices and techniques developed in this region for extraction of timber and conversion of logs have served as guides for related forest work in many other regions of the world. Such practices have been directed to the processing of both large and small diameter timber, but it was the former which dominated the discussions in Singapore.

From that broad stage-setting presentation by Mr. Carl Mason, the Singapore Seminar program progressed within parameters keyed to "local" or Southeast Asia resources.

Three fundamental markets for Southeast Asia hardwoods were analyzed: North America, Europe, and Japan, each being described by a speaker from the respective market.

The reader will find that after the basic resources and the market potentials are set forth the Seminar proceedings move directly into the operational and mechanical processes which are the means to achieving production and market goals. For example, the importance of reliably grading hardwood timber is covered, as are production methods for tropical hardwoods.

In fact, the key to the Singapore Seminar was the in-depth look at operational functions and methods—just as it has been in other Seminars and Clinics. As a result, the chapters of this volume range through production methods, portable sawmills for heavy hard-to-reach logs, utilization, mill layout, North American equipment for increased accuracy in production, use of edgers in reduction of waste, and generating immediate profit from unsophisticated machines.

Other topics covered are Best Opening Face (BOF), saw doctoring, tool and mill maintenance, dry kiln operation, briquetting of waste to generate steam or to make saleable products, and veneering and layup of tropical hardwoods.

Although the Singapore program was devoted almost totally to methods and markets, another topic—different but essential—was included. This was training programs for operating management and personnel. Whether a production facility utilizes unsophisticated equipment or the most modern machinery obtainable from factories in Japan, Europe, or North America, people are vital to successful, profitable operation of that equipment.

The initial Southeast Asia Sawmill Seminar did not seek to present encyclopedic coverage of conversion of the area's forest resources and the marketing of those resources. It did, however, seek to focus on the importance of those resources and the necessity for utilizing them efficiently.

Naturally, the formal talks were the pivotal point of this international gathering. But, as has been the case in Seminars and Clinics in Europe and the United States, informal discussion and conversation among registrants and program participants afforded invaluable additions to the formal talks.

The contents of this volume of proceedings, while directed to Southeast Asia operations and resources, will be of considerable value to readers who are interested in sawmilling outside the geographic area encompassed by the Singapore program.

Every effort was expended to assemble a speaker-list of persons extremely knowledgeable in their respective fields and familiar with Southeast Asia's forest resources and problems.

A number of persons besides the writer contributed to the project and I wish to recognize their work—primarily, Mr. Vernon S. White, of Miller Freeman Publications' Portland, Oregon, office, who served as liaison in contacting potential speakers and in selection of program topics. Many of the topics finally chosen resulted from replies to several thousand questionnaires mailed earlier to industry and government people in Southeast Asia.

Mr. James C. Wallace, publisher of *World Wood* magazine, visited Southeast Asia, as did the writer, to investigate the opinions and wishes of potential Seminar registrants and to explore different meeting locales.

Other Miller Freeman Publications editors contributed ideas based on experiences in Seminars or Clinics previously held in Munich, Germany; Jönköping, Sweden; Portland, Oregon, and New Orleans, Louisiana, USA. These included staff members of *World Wood* and *Forest Industries*.

This compilation of proceedings could not have been completed without the aid of the book department editing and production staffs and I wish to express our appreciation to them.

Finally, and most importantly, appreciation is expressed to those who were the heart of the Seminar: the registrants, the speakers, and the manufacturing firms who exhibited products or described available services.

Duane C. Mason
Associate Editor, *World Wood*
San Francisco, California
September 1975

H. C. "Carl" Mason, president and technical director of H. C. Mason & Associates, was raised in the Douglas fir logging and lumber community of Mapleton, Oregon, USA. In 1943, he entered the University of Oregon on an athletic scholarship, majoring in physics. Two years later, he joined the U.S. Navy, where he continued his eduction in electronic engineering. He returned to the lumber manufacturing industry as an engineer and contractor of sawmills and other wood conversion plants. In 1953, he joined Columbia Lumber Co. in Juneau, Alaska, as a consultant, and the following year acquired Fryer Corp. of Portland, Oregon. He then began to expand into the design, engineering, and construction of sawmills, veneer plants, and plywood plants. The growing importance and demand for economic feasibility work led him to establish H. C. Mason & Associates in 1963. His affiliations include memberships in the Institute of Electrical and Electronic Engineers, the Institute of Management Consultants, Association of Management Consultants, the Plywood Research Foundation, and the Southern Forest Products Association. He is also a graduate of the Advanced Management Institute of the Harvard Graduate School of Business Administration.

Robert F. Dwyer is vice president of Dwyer Overseas Timber Products Co., in Portland, Oregon, USA. He earned his bachelor's degree in business administration at the University of Notre Dame, and also studied at the Graduate School of Business, Stanford University. Dwyer is a third generation lumberman. His family has been active in Oregon in logging, lumber, and plywood since the early 1900s, and before that, his grandfather was a logger in Minnesota. Dwyer was active in the sales and administrative ends of his family business, the Dwyer Lumber & Plywood Co., during 1960-1964. His family operated a fully-integrated forest products complex manufacturing lumber, plywood, and a complete range of by-products. This firm was sold to a large, publicly-owned company, and, during 1967-1971, Dwyer served as president of the Dwyer Memorial Hospital. In 1969, he became vice president of Dwyer Overseas Timber Products Co., which imports tropical hardwoods into the United States and Canada.

J. L. Speechly is a senior director of Price, Morgan (Hardwoods), Ltd., which is the hardwood trading company within the Price & Pierce Group and one of the leading firms dealing with hardwoods in the United Kingdom. The Price & Pierce Group deals in all types of forest products from all parts of the world and is probably the largest in its field. The Group has branches throughout the world and has recently added to these by forming companies in Austria, West Germany, and the Netherlands. Apart from dealing in forest products, the Price & Pierce Group has a strong international transportation network and performs an increasingly important financing function in the United Kingdom and Europe. Speechly has many friends in the Far East and has also visited West Africa and a number of countries in Europe. His chapter is concerned with the market for tropical hardwoods in Europe. His outlook for the timber industry in general is optimistic, as he foresees a growing housing and furniture market. Speechly has little doubt that Southeast Asia will supply an ever-increasing proportion of the European hardwood requirements.

Mikio "Mike" Sasaki is a first-class registered architect and managing director of his own architectural and engineering firm, Sasaki Co., Ltd. He also serves as an assistant professor at Kogakuin University of Technology. Sasaki has published four books and many articles on architecture. His graduate publications at the Illinois Institute of Technology, Chicago, are rated among the 100 best works ever produced there and are on exhibit in several Far East countries. He has been awarded fellowships by Illinois Tech and by International House, Chicago, obtaining a master's degree in architecture from the former in 1964. Sasaki served as assistant vice president of the Tokyo Junior Chamber of Commerce in 1974. He was unable to attend the seminar in Singapore, but did prepare a paper entitled "The Market for Tropical Hardwoods—the Japanese Market" which was delivered at the seminar by Duane C. Mason, associate editor, *World Wood*. Sasaki's chapter concerns the Japanese market for tropical hardwoods. Considering the Japanese people's fondness for houses made of wood, Sasaki sees Japan as a great potential timber market.

H. C. "Carl" Mason

Robert F. Dwyer, Jr.

J. L. Speechly

Mikio "Mike" Sasaki

Edwin B. Huddleston is a consulting engineer with a long and varied work history in the Southeast Asian timber industry. He is a native of MacLean, New South Wales, Australia, and now resides in Sydney. His degrees include a bachelor of science as well as a bachelor of engineering in civil engineering. His experience includes five-and-a-half years as assistant research officer, Division of Wood Technology, Forestry Commission of New South Wales, and 30 years as officer in charge of that division. During World War II, he was licensed by the Defense Department and the Department of Civil Aviation for the inspection of timber plywood and glues, and he directed a large staff in these activities. Huddleston has undertaken missions in Malaya, Ceylon, and South Vietnam as a Colombo Plan expert, in Chile as a senior adviser on forest industries in an F.A.O. program, and in South Vietnam on behalf of the New South Wales Government. His affiliations include the Australian Institute of Building, the Institute of Wood Science, the American Wood Preservers Association, and the Royal Institute of Public Administration.

Mervyn Page is officer-in-charge, Forest Conversion Engineering Group of the Commonwealth Scientific & Industrial Research Organization, Victoria, Australia. He is a fellow of the Royal Melbourne Institute of Technology (civil engineering) and an associate of the Institute of Foresters of Australia. He is responsible for research, design, and development of forest conversion plants and technology in his present position. Page has 26 years of experience with forest-based industries, with extensive experience in conversion of tropical forests. He is also a consultant to the United Nations on forest industries for developing countries. He is the author of more than 30 publications relating to forestry, and has lectured in 10 countries. His chapter is concerned with production methods for tropical hardwoods. He points out that the behavior of Southeast Asian tropical hardwoods, both as logs during conversion and as timber in use, is quite markedly different from other hardwoods as well as softwoods. He particularly emphasizes those differences which affect basic sawmilling procedures in Southeast Asia and are of particular interest there.

Maurice C. Taylor is managing director of Southern Cross Engineering Co., Ltd., Christchurch, New Zealand. He is a native of England, and served as a war service officer with the British Merchant Navy. His experience includes 18 years as a project engineer with Humphreys & Glasgow, Ltd., London, England. His present company designs and develops sawmills in both softwood and hardwood regions, with an emphasis on the smaller units. Accordingly, Taylor's chapter deals with the four principles of sawmill management—training, utilization, development, and balance. He relates the difficulties he had in his early efforts to convince small sawmillers to change to new methods, and emphasizes the need for sawmillers to change to new methods when the old ones are superseded. Although he has lived and worked in large industrial countries, Taylor notes that he has learned, by living and working in New Zealand, the value of a small and independent country or business that can think and act for itself. Taylor suggests various ways for Southeast Asian governments and industry to work together in order to reduce reliance on foreign aid.

Bruce A. Wick is director of Wick Industries International, Vancouver, Washington. He is a native of Longview, Washington, USA, and studied business adminstration at Clark College, Vancouver, Washington, and the University of Portland, Portland, Oregon. He is the founder of Wick Industries International, and has been involved in the forest industry business in Southeast Asia for eight years. His experience includes the position of plant manager for Electro-Ray Manufacturing Co., Vancouver, Washington, a manufacturer of electric baseboard heaters. Wick emphasizes the need for portable sawmills for timber in heavily forested or hard-to-reach areas. He explains that portable sawmills are generally simple to operate and maintain, and can save considerable time, energy and money for the sawmiller, if properly used. Wick notes that by using portable, or mobile, sawmills, the sawmiller can save up to 60 percent of his log trucking costs. He maintains that cutting logs into cants or dimension lumber in the forest eliminates the wood waste problem, and provides a chance to recover good wood from cull logs, which normally are left to rot.

Edwin B. Huddleston

Mervyn Page

Maurice C. Taylor

Bruce A. Wick

Klaus H. Schmieder, managing director of T-Export GmbH Hamburg, West Germany, was born in Hannover, Germany. He has an extensive technical and commercial education, specializing in machinery and economy. He is a specialist in wood processing machinery, with many years experience in overseas sales of this type of machinery. Schmieder's company, T-Export, has the exclusive selling rights to JEVO sawmilling machinery, which has been produced for 90 years, and currently sells heavily in African and South American countries. Schmieder uses the experience gained by his company in Africa and South America to project the feasibility of using mobile sawmilling methods in the tropical hardwood forests of Southeast Asia. He lists the various types of mobile sawmills and emphasizes the technical and economical differences between the various choices. He refers to the problem of utilizing wastewood and explains that the semi-mobile, horizontally traveling bandsaws are the most economical and have proven useful in Southeast Asian forests.

Muhammad Sunaryo Hardjodarsono is director of the Forest Products Research Institute of Bogor, Indonesia. He is a native of Magelang, Central Java, and received his master of forestry degree at the University of Indonesia in 1955. He also holds a master of science degree from the New York State College of Forestry, where he specialized in the pulp and paper field. He has served as a lecturer in wood technology at the Bogor Institute of Agriculture, and his Indonesian forestry background also includes posts as the Chief, Bureau of Economics, Department of Forestry, and as the Director of Forest Utilization, Department of Forestry. He presents an overall look at the conditions of forest utilization in Indonesia, with an emphasis on current economic development in Indonesia. He notes that Indonesia has vast untapped timber resources, of which about 24 million hectares are economically accessible. These untapped resources are capable of producing up to 34 million cubic meters of logs annually, and Hardjodarsono emphasizes the necessity of continued improvement of timber resources management in Indonesia through private initiative supported by government pioneering.

H. A. "Harry" Hallewell is manager-equipment sales of MacMillan Jardine (M) Sdn. Bhd., Kuala Lumpur, Malaysia. He was born in Canora, Saskatchewan, Canada, and graduated from the University of Manitoba with a master of science degree in mechanical engineering in 1963. After graduation, he worked for two years with Consolidated Paper Corp. (now Consolidated Bathurst) in the research and development division, working primarily on groundwood for newsprint furnish. In 1965, he moved to MacMillan Bloedel, Ltd., Vancouver, B.C., Canada, where he worked in both lumber and shingle mills in positions of staff engineer, planning and development engineer, and, finally, manager of planning for the lumber and shingle group. He transferred from that position to a production job as processing superintendent in charge of planning, lumber handling, and shipping for a large MacMillan Bloedel mill in the Vancouver area. He transferred to his present position in January 1975. He is a member of the Association of Professional Engineers of British Columbia. His topic for the Singapore seminar, "Mill Layout for Maximum Efficiency," is one which he is well qualified, by both education and experience, to discuss.

Philip Kendrick is vice president and general manager of Canmillex Sawmill Exports, Ltd., in West Vancouver, British Columbia, Canada. He studied aeronautical and mechanical engineering in England at Cowes Technical Institute and Southampton University College. His company, Canmillex, is a consortium for export sales, including CAE machinery, Elworthy and Company, L-M Equipment, Mainland Foundry and Engineering, and others. Before joining Canmillex, he gained experience with the Canadian Sumner Iron Works (later renamed Black Clawson), where he served in the capacities of chief engineer, manager-European division, and director of sales (USA). He has also worked as general manager of Lamb Cargate Industries. He is a member of the Institute of Mechanical Engineers. Kendrick was unable to attend the seminar in Singapore, but did prepare the paper entitled "Canadian Sawmill Equipment to Increase Accuracy," which was delivered at the seminar by Adrian Kuypers, one of Kendrick's associates at Canmillex.

Klaus H. Schmieder

M. S. Hardjodarsono

H. A. "Harry" Hallewell

Philip Kendrick

Adrian J. Kuypers is a sales engineer with Canmillex Sawmill Exports, Ltd., in West Vancouver, British Columbia, Canada. He was born in the Netherlands, and earned his degree in mechanical engineering at the Institute of Technology-Amsterdam. His work experience includes positions with H. A. Simons International, Ltd., where he worked as a design engineer, and Columbia Engineering International, Ltd., where he served as a project engineer. Kuypers spoke to the seminar on the subject of "Canadian Sawmill Equipment to Increase Accuracy." In his speech (which was prepared by Philip Kendrick, vice president and general manager of Canmillex), Kuypers stresses the importance of well-designed machines to any sawmill. He also makes note of the importance of the operational personnel of the sawmill. He emphasizes that the head sawyer and saw filer are the most important people in the sawmill, and that their performance can strongly affect the attitudes of the other mill personnel. The sawmill arrangement discussed in the chapter is the carriage and bandmill type of headrig rather than the frame saw.

Robert Arciaga works in export sales for Canadian Car (Pacific) division of Hawker-Siddeley. A native of Manila, he studied business and marketing at the Philippines' State University, beginning in 1956, and took supplemental technical studies by correspondence while launching a marketing career in 1961. With eight years of field experience in varied lines, including wood machinery and supplies, to his credit, Arciaga joined the Canadian Embassy in Manila in 1969, as commercial officer specializing in agricultural and wood industry products and equipment. He emigrated to Canada in 1973 where he began intensive in-plant training with Canadian Car (Pacific). He now travels between Manila and Vancouver, assisting in the initial stages of Can-Car's expansion into Southeast Asia. He is a member of the Philippine Marketing Association. In his seminar topic, "Modern Technology in Lumber Processing," Arciaga concentrates on two related topics: the recovery of wood chips from sawmill waste, and recent developments in high-production machines. He also calls upon Southeast Asian sawmillers to diversify their products.

Mark Lawrence is president of Mark 50 Machinery, Portland, Oregon, USA. He is a native of Tillamook, Oregon, and was educated at the University of Oregon and Portland State College. His background includes sales experience with R. Hoe & Co., Portland, Oregon, and a vice president-sales position with M. A. Ward Corp., Eugene, Oregon. His present company, Mark 50 Machinery, manufactures a complete line of sawmill machinery from log decks to planers. His seminar topic, "Use of Edgers in Reducing Waste and Speeding Mill Flow," emphasizes the importance of the edger in the sawmill layout. Lawrence recognizes the fact that, in many Southeast Asian sawmills, the edger is not in current use, and he attempts to simplify and clarify the utilization of this machine as a basic unit in the sawmill layout. He notes that there is a waste potential in any cutting tool that is operated by people, and warns that an operator must be conscientious and well-trained to contribute to maximum recovery, even when using modern machinery such as the edger. He concludes that, although there are many variations of the edger process, it is the proper second step in most sawmill layouts.

Raymond C. Isles is managing director of Isles Forge and Engineering Pty., Ltd., Coffs Harbour, New South Wales, Australia, and is also a director of several other companies associated with forest products. He is affiliated with the Australian Institute of Production Engineers, the Australian Institute of Management, and the Australian Institute of Directors. He is the founder of his company, which is closely associated with sawmilling and timber processing in the design and manufacturing fields. His company has won many Australian Industrial Awards for good design, with special emphasis on export achievement. Isles has participated in Trade Missions to the Southeast Asian area, always representing the forest industry section of the mission. Isles Forge and Engineering Pty., Ltd., is the largest sawmill engineering organization in Southeast Asia, and the company has been associated with many major projects relating to the timber industry in the area. Isles emphasizes, in his seminar topic, the need for less sophisticated, or complicated, machinery in the area.

Adrian J. Kuypers

Robert Arciaga

Mark Lawrence

Raymond C. Isles

Gert Harloff is general manager of Vollmer (Singapore) Pte. Ltd., a leading manufacturer of saw maintenance equipment. A native of Hamburg, he has worked in the woodworking industry for more than ten years. He has practical training in sawmills, plywood and veneer factories, and furniture factories. His seminar topic, "Proper Saw Operations and Effective Saw Doctoring," deals with the element that all cutting tools have in common—a cutting edge similar in action and shape. Harloff explains that a successful sawmilling enterprise owes a great deal to the performance of the cutting edges of its saw blades, and that when paying attention to proper maintenance, even with a relatively small investment in tool maintenance machinery, the sawmiller will find success even with simple machinery. Harloff covers his topic thoroughly, listing and explaining the different types of machines for the maintenance of saw blades and the special features that each of these machines offers. He also gives important details on the process of hardening the tooth points of blades that will lengthen the service life of the machinery.

Erik W. Sundstroem is manager of long-range technical planning for Sandvik Steel in Sandviken, Sweden. A native of Karlstad, Sweden, his degrees include a master of engineering in engineering physics, and a doctor of science in applied mechanics. Before coming to Sandvik Steel in 1970, he had been associated with Bofors Steel since 1962. He has published articles on plastic deformation of metals and on lichens as indicators of pollution in forests. His special interests include tools and timber-decaying fungi. He is a member of the Swedish Technologists Association and the World Future Society. In his seminar topic, "The Economic Importance of Proper Tool Maintenance," he notes that it pays to use the best available tool and maintenance. He explains that the small tool budget has a decisive influence on the company's earnings, and that better tools or maintenance equipment is often the best investment the sawmiller can make to keep down rejects, downgradings, and waste. He warns against buying tool "bargains" as a method of saving money, because in the long run, this type of savings will prove to be a loss.

Clair A. Smith is marketing manager of Filer & Stowell, Milwaukee, Wisconsin, USA. A native of Coatesville, Pennsylvania, USA, he holds a master's degree in metallurgy from Pennsylvania State University, as well as a master's degree in marketing from Kent State University. His professional background includes positions as sales/marketing manager for Medalist Industries, national sales manager for Trans Union Co., and district sales manager for Lukens Steel Company. He is a member of the advisory board of the Southern Forest Products Association. In his seminar topic, Smith notes that no matter how accurate a piece of equipment might be, it can only perform to its claims if the mill is properly aligned. He then discusses the equipment used for aligning the mill, confining his methods to materials that are most likely to be found in any sawmill—such as wire, center punches, "C" clamps, etc. Smith also details the process of aligning each piece of mill equipment, starting with the "V" rail, and then proceeding to both the edger and the trimmer.

Ali Zafar is managing director of Hildebrand Singapore Pte. Ltd. in Singapore. A native of Calcutta, India, he received his bachelor of engineering degree from the University of Karachi, and his master of engineering degree from the Asian Institute of Technology. He has worked as senior engineer in engineering design for Frederic R. Harris, Inc., New York, and in the same position for Doxiadios Associates, Athens. Since 1970, Zafar has worked in Southeast Asia for the Hildebrand company, first as the Far East delegate of the company, and later as managing director when the German company was incorporated in Singapore in 1973 with the aim of local production of timber-drying equipment. Other Hildebrand subsidiaries operate currently in Spain, Sweden, Japan, France, and the United States. His seminar topic, "Proper Dry Kiln Operation," emphasizes that kiln-drying of sawn timber is one operation that helps standardize the quality of a sawmill production; thereby obtaining the trust of the buyer for a long-lasting trade relationship. Zafar also notes that there is a ready and established demand for kiln-dried lumber; therefore, it commands premium prices.

Gert Harloff

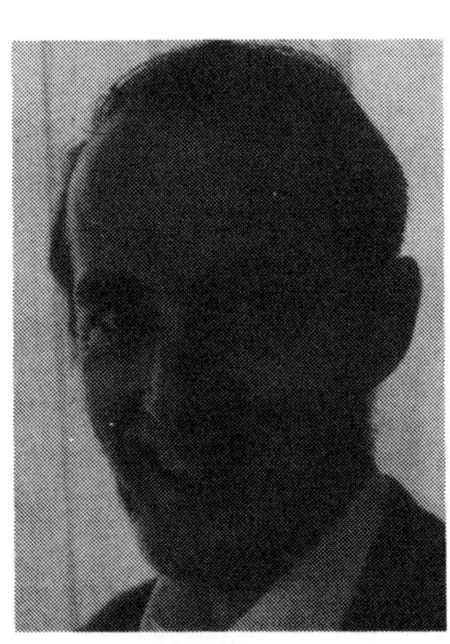

Erik W. Sundstroem

Clair A. Smith

Ali Zafar

Fred Hausmann is a native of Switzerland and president of Fred Hausmann, Ltd., formerly Glomera Holding Ltd., of Basel, Switzerland. For the past 20 years, his overwhelming concern with what he saw as the potential productivity of industrial wastes caused him to concentrate on the problem of waste utilization. In these wastes, Hausmann could see huge amounts of potential energy lying unproductive. This led him to invent briquetting devices which convert a large variety of residue materials—such as bark, sawdust, pulp fiber, almond hulls, peanut shells, aluminum and copper swarf, etc.—into useful sources of fuel. Marketing these devices has taken him around the globe. A physical fitness enthusiast, he skis cross-country, swims, water skis, climbs mountains, plays tennis, and combines his love of the outdoors with photography. In his seminar topic, Hasumann points out that much financial success—including that of John D. Rockefeller—is based on recovery and use of waste. He concludes that 21 million tons (dry weight) of unused bark and other wood wastes generated annually in Alaska, Washington, Oregon, and California could be converted into 11 million kilowatts of electric energy.

Tim A. Bailey is general manager of Jurong Plywood Co. (Pte.) Ltd., an affiliate of Boise Cascade, in Singapore. A native of Colorado, USA, he received a bachelor's degree in accountancy from the University of Denver. He has been employed by Boise Cascade since 1965 in various financial positions including manager of finance, Southeast Asia. Bailey is chairman of the Singapore Plywood and Veneer Manufacturers' Association and a member of the Singapore Timber Industry Board. His seminar topic, "Veneering and Layup of Tropical Hardwoods," details the steps necessary in building a complete plywood mill and selecting the proper equipment for it. Bailey warns against building plywood mills on too large a scale because of improper planning. He states that there are many Southeast Asian plywood mills with surplus equipment that has never been used, or that has been used and found totally unsatisfactory. In addition to proper equipment, Bailey emphasizes the importance of considering the plant shape, the wood sources, and the panel mix in the design of a plywood mill.

Leonard H. Taylor is a consulting engineer with C. D. Schultz & Co., Vancouver, British Columbia, Canada. He earned a bachelor's degree in mechanical engineering at the University of British Columbia. He has worked in several industries in several countries. Taylor's background includes engineering positions with Khulna Newsprint Mills, East Pakistan; Power House Construction, Jamaica; and Standard Oil in Venezuela and Iran. Much of his work has been with sawmill and plywood plant design, construction, and operations. He has also been the project manager for the construction of two large pulp mill complexes in Canada, and so he is well acquainted with many of the problems of the timber industry. He is a member of the Association of Professional Engineers in British Columbia. An outdoorsman, Taylor has worked three winters as a professional skier, and is also an avid golfer and scuba diver. His seminar topic, "The Importance of Good Training of Sawmill Workers," emphasizes the importance of adequate and continuous training of all operations workers and of members of the staff of any sawmill.

Fred Hausmann

Tim A. Bailey

Leonard H. Taylor

CHAPTER 1

Matching Southeast Asia's Wood Resources to its Opportunities

H. C. "CARL" MASON
President, H. C. Mason & Associates, Inc.
Gladstone, Oregon, USA

In this chapter we will be on a journey together. That journey is being traveled by each of us, as individuals, in a quest for ways to create opportunities to use the vast wood resources of the many countries in Southeast Asia, a vital part of the world community. Our journey also is being pursued to discover means for the development of proper wood products industries for each of the countries in that area.

From its outset, our journey can be seen as a long one, and the path a difficult one to follow. Together, we are seeking new ways of utilizing the indigenous wood resources of a geographic area extending more than 4,000 miles while spanning two oceans and many seas. Within this region, too, live more than 230 million persons speaking different languages and literally hundreds of dialects.

At first glance, even the variations of wood species are staggering to comprehend in an orderly way because of their complexity and number. Species range from the hardwood forests of tropical and subtropical areas of the Philippines, Malaysia, Indonesia, Burma, Thailand, and New Guinea to the plantation pine forests of Australia, New Zealand, and the Fiji Islands.

Therefore, because of the many differences of wood species, among people, in languages, and in topography, it may, at first impression, seem impossible to believe that we can all travel along the same route together. Yet, I believe we can, for we are all united in the pursuit of one common objective: that objective is to improve our knowledge about the profitable conversion of wood into its myriad marketable products.

If I can offer you any encouragement, it is to remember that you will succeed if you seek more than knowledge about methods, as important as that is. You will succeed if you look beyond the workings of the fine machines, so vital to our industry. You will succeed if, instead, you try to discover the secret of utilizing both methods and machines and to see them as part of a total system.

Conversion process

To understand the philosophy of a modern wood products technology, we must come to think of a conversion process as a single system, but one composed of five interrelated and indivisible elements. These elements are: resources, capital, operational technology, personnel expertise, and market opportunities. Together, they compose the systems technology vital to a profitable wood products industry.

One of the many advantages of my own profession, that of a consultant, is that I have had innumerable opportunities to travel around the world and to view firsthand the state of the art of wood conversion. From my experiences, I have discovered one principle which I believe is universal to all countries engaged in or planning to develop viable wood products industries. That principle is as follows: in the conversion of wood into its many products, every effort must be taken at every step in the conversion process to save fiber. In the planning of all conversion systems the goal must be to optimize fiber recovery.

Contrary to some people, when I view the relatively low level of technology to be found in the emerging nations of the earth, I do not despair. That is because the concept of fiber recovery is not thoroughly understood and accepted by the wood products industry in even the most technologically advanced nations. Accordingly, there is much to be optimistic about in those countries now developing wood products industries of their own for they will be able to develop systems that implement fiber recovery.

Figure 1.1 is a histogram which shows the relationship between log diameter and recovery in the sawing of logs into lumber products. It is evident that recovery in small boards is difficult to improve upon in small logs. In larger logs, however, as the figure shows, the potential for additional lumber recovery grows. It should also be noted that with logs as small as 12 to 15 inches, recovery rates are significantly high.

Accordingly, it is in the recovery of fiber that there is to be found the greatest opportunity for profits and the recovery of capital within the lumber industry, as it now exists. Simply stated, there is a great opportunity available in the conversion of unused fiber into salable products at competitive prices. Yet, in order to attract the capital which is available, and which is so badly needed in Southeast Asia, planning and organization of new capital structures must be directed toward this opportunity.

Accordingly, the planning in the wood conversion process required to exploit this great opportunity must extend from the standing forest all the way to the world markets. This planning has become increasingly complex to a point where, in North America today, it is relatively common to do this work arithmetically with computers. There are econometric modeling programs by which we can evaluate the net result of different end-product distributions projected from incremental differences in log diameters and lengths. These projections determine a product spectrum which will produce maximum values.

A practical application of these planning methods can be clearly understood in the area of energy cost rate equivalents between the alternative power sources. Throughout the world today, the cost of energy plays a most important part, if not the most critical role, in planning capital expenditures to seize profit opportunities in the production of wood products.

In those areas of the world where wood industries are well established and relatively advanced in their technology, the basic planning problem is to devise means to reduce or totally eliminate a dependency on petroleum as a primary energy source. Yet in those countries striving to develop a wood industry, this kind of thinking may not be applicable at this time.

Sound planning

The problems inherent in the development of a forest industry through promotion of small, independent sawmills are poor or wasteful use of the resource; difficulties in maintaining uniformity and quality in finished products; and lack of a consistent output of identifiable finished products. Generally, lack of capital is the reason for this. Obviously where capital, raw material, and labor are available, the most efficient and most profitable vehicle would be a large, integrated central manufacturing facility. A commitment to such a facility ultimately becomes a social and political problem as much as an economic problem, depending on the goals of a government in planning for a forest products industry.

In other words, what is sound planning in one part of the world is not applicable to another. In developing a wood industry from a relatively low level of technology, people have to start somewhere. Accordingly, a small diesel generator-type sawmill operation is often better than no mill at all. Such installations offer the beginnings for capital accumulation. They are a place from which to grow with a minimum investment and relatively low overhead costs. Although actual economic data from such installations are hard to obtain, it is, assuredly, subject to the same economic analyses procedures which should be the foundation of all planning.

The ability to plan purposefully, for a specific market and with a specific raw material in mind, is probably the largest single area of performance improvement available to any lumber manufacturing operator today, anywhere in the world, regardless of species.

Having formulated a viable plan for future conversion, the selection of the production systems to make a plan a success is the next step. Yet the value of any single system or machine can be proved only if its usage is consistent with the overall systems technology determined by the master plan. I refer specifically to the uniformity, or the mechanical excellence of the system. Provided there is a plan and a specific purpose for the breakdown of a log into lumber products, the machining system can follow this plan with a high degree of accuracy.

Throughout the world, it is not uncommon to find lumber variation in excess of 1/8 of an inch—and in many cases 1/4-inch—especially on rough lumber conversion with simple head saws and carriages. Figure 1.2 shows the relative rates of recovery and expected improvement in recovery by saving a few thousandths of an inch in the manufacturing process of the average log.

Obviously, accuracy and precision performance is important and there is a direct relationship between the amount of wood consumed and the amount of products produced. When we talk of systems to improve this— and improvements include increased machine accuracy, thin kerf saws, drying systems that have less shrinkage, and smaller planer allowances—we are referring directly to the machining system.

Closely allied to this factor is the area of operations, which is the management of the total operational plan, including the machining system. Ultimately the success of the operational plan rests solely on the performance of individuals—the labor force—and how well they perform their jobs. I referred to this element earlier as personnel expertise.

Understanding the work force

If there is one opportunity which presents the greatest potential for improvement in the world's wood products industry, it is in the area of improved performances by individual workers. Although this opportunity is available to all managers throughout the world, it is one which is frequently overlooked.

To take advantage of this opportunity, however, it is necessary to begin with an inventory of the skill levels required and those available. In connection with a reference paper prepared at the request of the United Nations Food and Agriculture Organization, we have developed a series of job classifications for the wood industry.

Figure 1.3 shows these job classifications broken down into group, job title, work performance description, and skill level and percentage required to have the skill level. Skill levels run from one through ten, with a one-level skill being the lowest rating.

In order to obtain a perspective on a work force, it is important to note that 31 percent of the labor force is at level eight or above. This means that such employees will require professional-technical training, or the equivalent experience, before they come on the job. They cannot be trained after hiring. Another 16 percent of the labor force can be, conversely, put right to work and learn in a matter of hours what is required of them. These persons are in the first three skill levels.

Finally, a significantly large number, amounting to 53 percent, of the jobs are in levels four through seven. These people can be trained after hiring, provided there is an extensive on-the-job training program given them. These are the people who will require the most attention in the development of a skilled work force. But it is significant to note that well over half of the jobs in the typical wood products facility can be held by persons trained to perform them without having experience or backgrounds *before* being hired.

Accordingly, the substance of this opportunity to improve labor force performance and, thereby, the implementation of operational planning, lies in management's being aware of which skills are necessary and in being committed to help train the indigenous labor force to fill vital jobs.

In most cases in the world, we have found that a return on investment can be excellent, provided that the people charged with the operation of equipment are committed to a high level of operational performance and to a quality control level necessary to make best use of their equipment.

Therefore, anyone who endeavors to improve upon the profitability and performance of a wood processing operation must be extremely concerned with the development of personnel communications, including training; careful maintenance; and a criteria for quality control. These efforts in most existing operations certainly will result in improved wood recovery and vast profit dollars without the necessity of any capital investment.

Marketing opportunities

Let me turn now to the subject of marketing opportunities. There are, throughout Southeast Asia, complex forests made up of a variety of species, all with infinite use possibilities, and this means additional opportunities for marketing. To be sure, many of the wood products in this area have been known and in demand in world markets for years. Yet to increase their share of the world's market for wood fiber, Southeast Asians must make—or create—an additional opportunity by accepting a formidable challenge. That challenge consists of producing a quality, uniform product acceptable and readily identifiable in the world's marketplaces. Such products will command the highest prices.

We believe that one of the greatest opportunities in dealing with southeastern Asian hardwoods, for example, lies in the abandoning of individual botanical species classifications. These classifications should be replaced, instead, with the use groupings of species. These species, accordingly, can be marketed in groups which are associated with both use and appearance. By the acceptance of use groupings and marketing products in this way, many minor species groups, heretofore virtually unmarketable, also can be included, because they will be quickly identifiable by purchasers. We have seen some recent, similar developments in marketing practices in the United States along these lines which have proved very successful by breaking into captive markets, long held by producers in various parts of the world.

Although opportunities to gain additional increments of world markets are great, the industry in SEA must not lose sight of opportunities for developing a strong domestic marketplace as well. Development of a strong local marketplace also affords an additional opportunity to develop vertical integration of this area's wood products facilities. Full integration will further open the way for maximum utilization of Southeast Asian forests, which is the ultimate goal sought by all nations having wood products industries.

In conclusion, let me say that the forests of Southeast Asia are only different from those of other areas of the world in some very positive ways. They are of lower cost, have higher intrinsic values, and afford more opportunity for operational profits than in most other wood-producing regions of the world. The problem is matching wood resources of Southeast Asia to the best market opportunities.

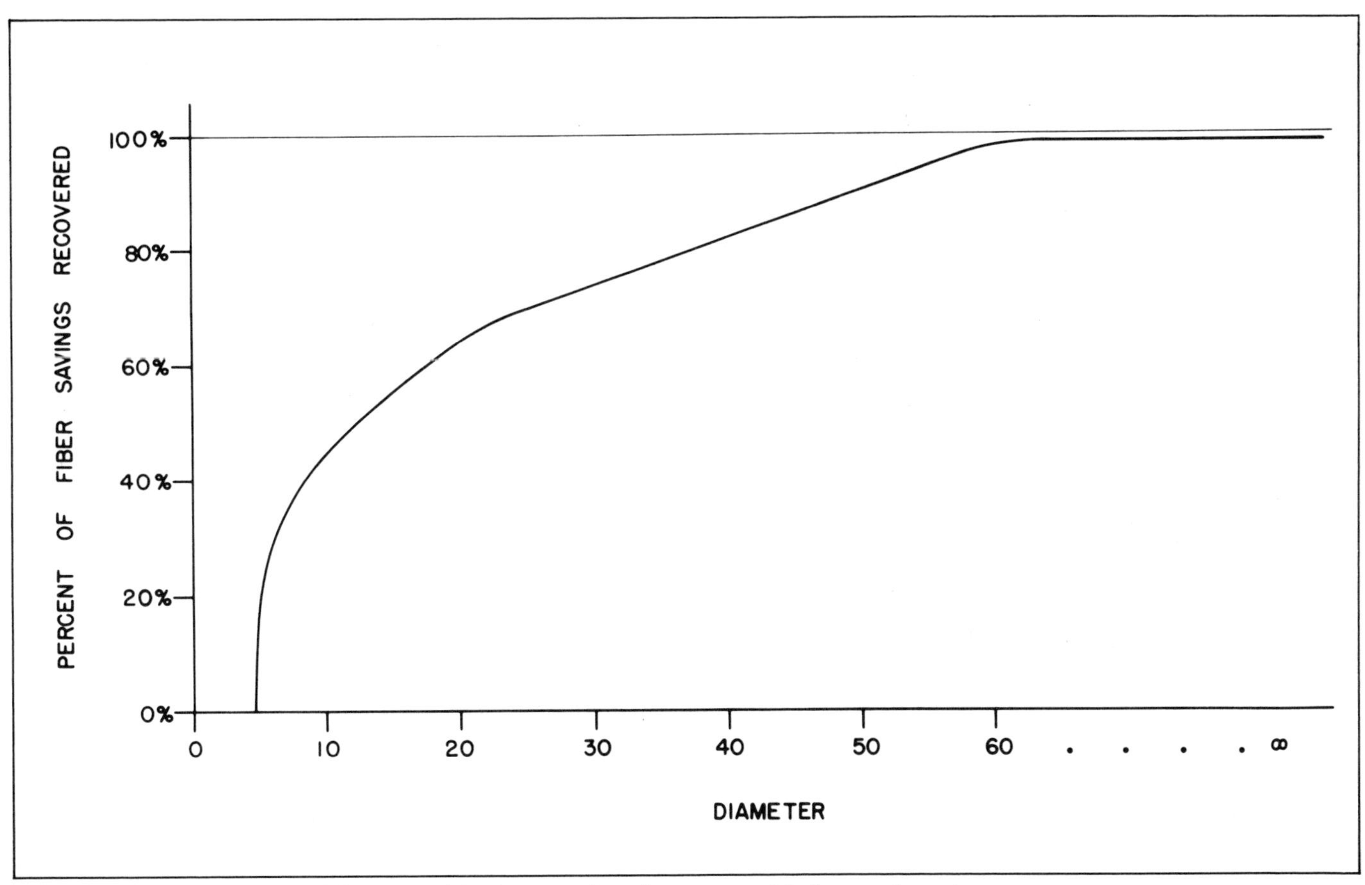

Figure 1.1. Percent of wood fiber saved by improved sawing recoverable into end product.

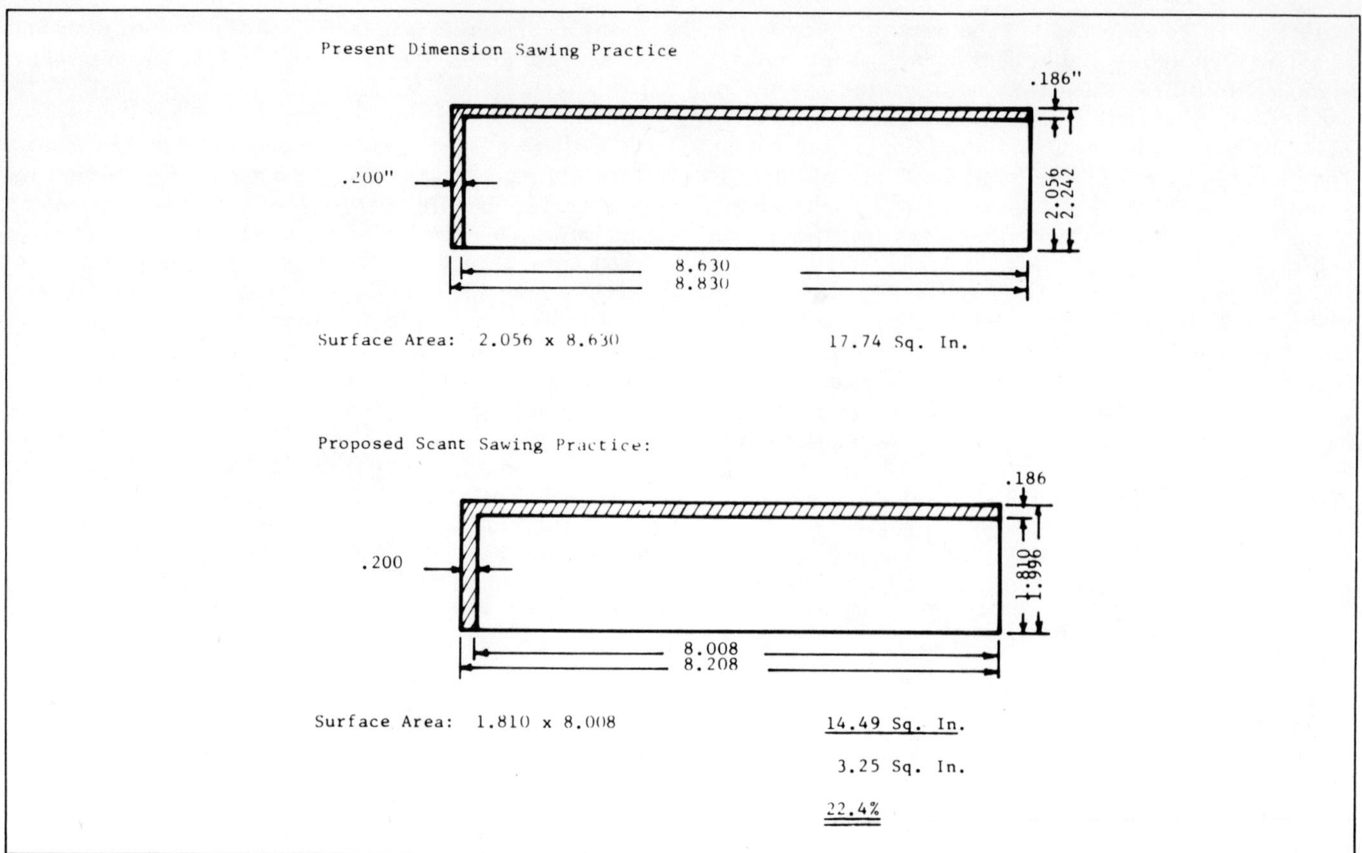

Figure 1.2.   Potential saving due to scant sawing (2″ x 8″).

Group	Job Title	Work Performed	Skill Level	% of Group
Manufacturing	Machine Operator No. 3	Operates flow directing machinery like drop gates, sorters, log decks, veneer trays or reels, patch equipment, presses.	4	9
	Equipment Operator	Operates lift trucks, straddle trucks, log handlers, cranes, etc.	5	8
	Grader, Certified	Grades material requiring a final shipping grade.	8	1
	Grader	Grades material requiring a grade only for further processing.	6	4
	Laborer No. 1	Tends machines like waste hogs and chippers, transfer chains, chip bins, straighten boards.	3	5
	Laborer No. 2	General labor, pulls veneer and lumber from sorting chains, feeds dryers and sorters, unloads presses, feeds sizing saws and sanders. Loads lumber and other products.	2	37
	Common Labor	Cleanup, floor sweeping, dust blowing, edging picking, bark scraping, waste hauling, fire tending, etc.	1	6
	Power Engineer	Operates power house and power generating equipment, air compressors blowers and pumps.	8	2
	Plant Engineer	Manages maintenance of plant equipment, designs tools and jigs, directs repairs where design is a factor, plans and designs plant improvements.	10	0.6

Figure 1.3.   Job classifications for a forest-based industry.

CHAPTER 2

The Market for Tropical Hardwoods—The North American Market

ROBERT F. DWYER, JR.
Vice President, Dwyer Overseas Timber Products Co.
Portland, Oregon, USA

What is the greatest barrier to exporting forest products to North America? Why does the Southeast Asian forest industry have problems in selling its products to North America? Why aren't there more orders and repeat orders from North America? The answer is simple: there is a lack of understanding on the part of the Asian sawmill operators about the North American market. Fortunately, this lack of understanding is a problem that is correctable.

I will use the term "North American" market to refer to both Canada and the United States. As far as wood products are concerned, both markets have more similarities than differences. Both have similar standards of living, similar market tastes, and similar business practices. There are differences, to be sure, but they are not as great as, for example, the differences between various countries within the European market.

In evaluating the North American market for wood products, we must at the very beginning recognize that North America possesses one very unique factor that makes its markets different from all the other major world markets. The Southeast Asian forest industry must recognize this difference and conduct its affairs accordingly if efforts in North America are expected to be successful. If these differences are ignored, then the efforts will not succeed.

This one basic factor which causes the North American forest products market to be different from the others is simply this: in addition to being the world's largest market for forest products, North America is also the world's largest producer of forest products. The United States, by itself, ranks second only to the Soviet Union in the production of lumber. The United States and Canada combined form the largest and greatest forest products industry in the world. No other major forest products market is as self-sufficient as is North America.

Because North America has this large and dynamic forest products industry, the market for these products has developed some unique requirements and characteristics. These can be best summed up by the use of the following words: quality, competitiveness, and diversity. The most important of these is quality.

The North American market is a quality market. The word "quality," as it applies to this market, is defined as "the degree of excellence which a thing possesses, hence excellence, or superiority." When I say the North American market is a quality market, I mean that only a product with inherent excellence or superiority will sell well in North America. Quality is the single most important requirement of the North American market, and it is mandatory, not optional.

We in North America enjoy the blessings of the free enterprise system. This means that we are free to use our own resourcefulness to try to do a better job than our competitor in order to get the business and make a profit. Notice that I said "better," not "cheaper." Doing a better job rather than a cheaper job is the proven route to success in North America. In this respect then, success depends on one's attitude. The successful businessman in North America has the attitude: how can I do it *better* and still sell it at a competitive price? How *good* a job can I do and still make a profit? An attitude typically found in those firms that are not successful in selling to North America is: how cheaply and poorly can I do the job and still get by with it? Because of our free enterprise system, American manufacturers use higher quality as well as lower prices as techniques to obtain profitable business.

In talking about quality, I want to stress one point: quality should not be confused with sophistication. Quality means the inherent excellence in what is offered in a given price range and market. Sophistication means highly refined or complex or lacking in simplicity. Consider, for example, two well-known German automobiles—the Volkswagen and the Mercedes. Both are quality cars. But the Volkswagen is a relatively simple basic car, whereas the Mercedes can properly be called a sophisticated car. If I told you that only a quality car will sell in the United States, and you thought that by quality I meant sophistication, then you would try to sell only Mercedes in the mistaken belief that there was no market for a Volkswagen. However, you would be wrong, because the Volkswagen sells very well in North America. And the reason it sells well is because it is a quality car for the market it is serving. Similarly, the North American market has room for forest products of every degree of sophistication, from rough, air-dried sawn-timber to the most intricate of millwork items and furniture parts. But the requirement for quality remains constant.

Quality also refers to your basic business dealings and means simply: always do what you say you will do and deliver what you contract for. Do not contract for one thing and then deliver something different. Anyone can do that. The North American customer does not have to come 9,000 miles just to get something he did not order; there are plenty of people right at home who are willing to do that for him.

The North American market is competitive—very competitive. It is competitive first of all because we have our own wood processing industry that can do many of the same things that Southeast Asian woods can do. For example, doorjambs are made of both our pine and Southeast Asian lauan; grapestakes come from Southeast Asian keruing and our redwood and Douglas fir; our oak and fir compete with Southeast Asian keruing and apitong for the railroad sleeper business; and bevel siding is made of our western red cedar and Southeast Asian lauan and meranti. Also, other wood-producing regions like Latin America and Africa are competing for this same business.

The market is also competitive because many substitute materials are readily available to do the same job. The suppliers of plastics, steel products, concrete products, and fiber products are all in there competing for the same business. If the price of wood gets too high or the supply becomes unreliable, another material will get the business. Some specific examples: mouldings are made of wood, wood fiber, and plastic; car decking is made of wood and steel; grapestakes are made of wood, steel, and plastic. Our customers are often results-oriented, which means they don't really care what material an item is made of as long as it does the job. As a result, our marketplace is a scene where literally thousands of resourceful businessmen are competing with each other to give the customer the best quality for the cheapest price.

Because of these factors—a large domestic production and a free enterprise system—the customer has many alternatives. You can only sell him what he wants to buy. This is in sharp contrast to other markets which do not enjoy their own timber reserves. For them, there are fewer alternatives. They are more dependent upon outside suppliers, and anything offered is better than nothing at all. But in North America it is different.

We have an expression in the United States that pretty well sums this all up: "The customer is always right." And he is. The American customer can truthfully say: "Give me what I want, when I want it, because if you don't, someone else will."

The North American market is a diverse market. It uses a lot of wood for a lot of purposes. It has a long and favorable heritage of using wood, and it is willing to try anything new that seems to make sense. This heritage of using wood dates back over 300 years when the first European settlers arrived in North America. At that time the trees in North America were as thick as they are today in Kalimantan. These earlier settlers used wood in hundreds of ways just to survive. This same resourceful and open-minded attitude towards the use of wood still prevails today. This means that there is a place for the Southeast Asian forest industry in our market. We have a million and one uses for wood, and a strange name or foreign origin is no barrier to its use in North America. Anybody who is resourceful enough to log and operate a mill in Asian jungles is certainly resourceful enough to make a quality product suitable for the American market. The market is diverse. It is big. It is getting bigger. The volume of wood being used and being imported is growing each year. The opportunities are limited only by one's own personal imagination and resources.

So now that we have talked about the general requirements which must be understood to successfully sell in North America, let's talk about specific product items. The first and most basic product is logs. Because of economic considerations and due to various governmental restrictions in the log-producing regions, it does not appear that logs are a good item for export to North America. There are exceptions, but I feel it is generally safe to conclude that North America is not a good market for Southeast Asian logs, and probably never will be. For centuries man has been arguing about whether the log should be processed at the stump or in the marketplace. Today's economic factors indicate that logs can be manufactured into products in the country of origin more cheaply than they can in the destination country, at least as far as North America is concerned.

Sawn timber or lumber

In the shorea species, the merantis and lauans, there is

a fairly steady demand for rough lumber. However, none of it is used in its rough state. Nearly all of it goes into a factory of some type which further processes the rough lumber into a finished product. Thus, the rough lumber is merely raw material for somebody else's factory. These factories are using the shorea species to manufacture such things as mouldings, door parts, jambs and casings, cabinet and furniture parts, and boat parts. In many instances the factory which uses lauan or meranti will also be making the identical products out of other species of wood. Sometimes these other species are used as an alternative when Southeast Asian woods become too expensive or scarce. But in other cases these factories regularly schedule the simultaneous use of several different species in order to give the customer just the exact item he wants. Because so many of the factories can and do use different species, one producer's sawn timber is constantly being compared with that of other producers. For example, if your sawn timber is cut more scant than is someone else's, the user will quickly note this. If your 4/4 will not completely surface to 3/4 of an inch, but another mill's will, then he will place a lesser value on your wood.

For the longer term, the prospects for selling rough sawn timber in North America do not look particularly favorable. The total footage may remain constant, but the amount of rough lumber as a percentage of all wood products shipped will diminish every year. There are two reasons for this. The first is that it is generally cheaper to make the finished end product in Southeast Asia rather than in North America. But the second reason is really the important reason. And that is simply that the quality of the rough lumber that is normally shipped from Southeast Asia is not acceptable to the North American customer. Too many customers have had too many bad experiences, and they have made the decision to avoid buying rough lumber whenever possible. Basically, the material is not up to the grade it is represented to be.

I have compared experiences with a number of my colleagues in the industry, and a general consensus seems to be that approximately 20 percent of each parcel of rough lumber shipped from Singapore and Malaysia to North America is not up to grade. This number—20 percent—continually comes up in conversations with others in the trade. Unless and until this situation is corrected, our customers will do whatever they can to avoid buying rough lumber. Because of these bad experiences, the prospects for selling rough-sawn lumber do not look good. I know that our firm, for one, has made the decision that it will not buy and pay for any lumber until it has been received in the United States and has been inspected by a certified inspector of the National Hardwood Lumber Association. This is the only assurance we have that we are actually getting what we ordered.

But let's take a positive and constructive note on this subject. Suppose, for example, that you are just building a mill in Southeast Asia and do not have the ability to make anything other than rough-sawn lumber. Can you expect to find a market for your production in North America? The answer is yes! And here's how to go about it.

First of all, you start by realizing that only your high grades of lumber are marketable in North America. We have more middle and low grades than we are able to use. We do not need to buy any from you. I told this to one Southeast Asian sawmill operator, and he said "Yes, but if I ship you only the high grade that you ordered, then what am I to do with the low grade?" He asked a very good question, and unfortunately I do not have an answer to this question. However, I will state with certainty that my customers cannot use low-grade lumber at high-grade prices, and will refuse to pay me for anything that is not up to the grade specified in the contract.

Secondly, I would tighten up on your log buying and log handling practices. It is obvious that you must have the right logs to fill your order file. But how many times has a mill accepted an order only to later tell the customer that he could not fill the order because he did not have the logs? And more importantly, how many defects, particularly borer holes, develop in the log between the time the tree is fallen and the log goes into the mill. As far as the grading rules are concerned, a borer hole is the same as a knot. But consider this: it took God 50 or 100 years to put that knot in the log, but some logger allowed that borer hole to develop in only one or two weeks!

Thirdly, manufacture your lumber better. By this I mean cut it straight, eliminate warp, trim the ends, develop uniform size standards and manufacturing tolerances, and stick to them.

I would suggest you obtain copies of the grading rules of the Western Wood Products Association, the West Coast Lumber Inspection Bureau, and the National Hardwood Lumber Association. The first two agencies are headquartered in Portland, Oregon, and publish the rules for the manufacture and grading of softwood lumber. These standards include size and warp tolerances for softwood lumber. The NHLA is headquartered in Chicago and publishes the rules for hardwood lumber. Although these rules may not apply exactly to your specific situation, all three rulebooks tell you what the American customer expects as regards quality of manufacture and the final size he expects your lumber to surface out to.

Next I would recommend that you install dry kilns. Since all of this lumber is dried before being used, why not install your own kilns and put the revenue from kiln drying into your own pockets?

Find out exactly what sizes your particular customers like best. Some like the wide boards the best. Consider a program of cutting each log to the widest possible sizes, regardless of maximum size. Other customers prefer specified widths, usually 6-inch, 8-inch, 10-inch, and 12-inch. Others prefer specified lengths. Perhaps the worst thing to make is straight random-width, random-length lumber. Every little gyppo mill in the world is doing this. If you do this also, then you will find yourself competing down on the same level, and it's a very difficult way to make money.

Most lumber in North America is sold in railcar-load quantities. Each car contains between 25,000 and 40,000 board feet. When you get an order for these quantities, be sure to ship the orders in complete quantities. The rail freight rates in the United States and Canada are based on the cars being fully loaded. The freight rate per 1,000 board feet increases drastically if the car is not full. When you ship an order incomplete, you may very well cause a large increase in freight costs because you did not ship enough to fill a car. This extra cost is a needless expense that does not do you or me or our customers any good—it only goes to the railroads. Therefore, it is important that you ship your orders in the specified quantities.

Lastly, and most importantly, ship your lumber to the grade specified. Learn your grading rules and understand them. If you are building a new mill, I suggest you hire an outside consultant to teach your men how to grade. In North America, all mills manufacture their lumber or plywood to published standards. They live by their standards and do an outstanding job of it. The American customer expects exactly the same thing from you, and if you cannot meet his expectations he will either turn to someone else who will, or will discount your price commensurate with the grade actually shipped.

Car decking

The North American market for truck and rail car decking and related items took a severe downturn in 1974 due to the recession, but there are some signs that it will pick up again in the first quarter of 1976. Several important points should be stressed regarding these products.

First of all, the points I just covered regarding the manufacture of rough-sawn meranti and lauan also apply to the manufacture of rough-sawn keruing and apitong.

Secondly, the number of uses in North America for sawn timber of these species is much more limited than it is for the shorea species. New uses are being developed, but at a slower rate. The principal use seems to be for decking for trucks and railroad cars and related uses as components of heavy equipment.

Although several manufacturers are successfully manufacturing decking in Asia, there will always be a strong requirement that much of it be processed in North America, for several reasons. First of all, the lead time between placing of the order and need to use it is often too short to allow placing an order overseas. Secondly, the specifications vary from one truck or railcar manufacturer to another. The suppliers of this decking need to keep blanks in stock that can be exactly machined to each customer's specifications as the orders come in. And thirdly, the railcar and truck manufacturers insist that an inventory be kept in the country. They are running big assembly line type operations and cannot afford to shut down their factories just because a few hundred dollars worth of wood is tied up on a boat someplace.

As in anything else, the market is very competitive. Car decking can be very competitive. It is good and fairly steady business, but you have to have a good price and be a reliable supplier to be in it successfully.

Railroad ties

The market for sleepers, or ties as we call them, appears to be limited in North America as far as the Southeast Asian forest industry is concerned. Our domestic industry supplies both oak, tupelo, and gum ties for the hardwood tie market and several softwood species for the softwood tie user. The ocean freight on ties is a significant part of the delivered price of Southeast Asian sleepers. As the price goes up, more and more domestic mills will start making them. There have been several sales of Asian ties into the United States and Canada, but these have usually been special situations. Our domestic timber reserves have large volumes of timber that will make good ties, and therefore I think that, as a practical matter, the Southeast Asian forest industry can forget about doing much sleeper or tie business in North America.

Grapestakes

The grapestake market is dead. It is not yet buried but it is dead nevertheless. Grapestakes are purchased only for new plantings of new vineyards. In 1972 and 1973 there were over 60,000 acres planted each year. Last year that fell off to 27,000 acres, and this year there will be less than 9,000 acres. The planting of vineyards was a boom that has gone bust. It is estimated that there is now enough acreage planted to sustain the wine industry in California until at least 1980.

The new grapestakes being ordered must be of all keruing. No kempas is permitted. We have learned that kempas stakes split when they dry out and thus become worthless. Also, the new stakes will have to be pressure-treated to .5 pounds per cubic foot on the oxide basis, which converts to approximately .9 pounds on the gauge method using British standards. A grapestake is now a precision product, and will be sold only in limited quantities by those who are prepared to do the proper job. However, it should still be a good recovery item for a sawmill having low labor costs.

Surfaced lumber

Lumber can be graded according to several different criteria; one is according to strength, as we do with our construction framing lumber. Another is grading on a cutting basis, as is "select and better" in the Malayan grading rules and "Firsts and Seconds" (FAS) in the NHLA rules. However, lumber can also be graded purely on an appearance basis. Appearance grade lumber, which we call "finish" lumber, is very widely used in North America. It is made from a number of softwood species. Basically it is completely free from defects and the entire board is useable in one solid piece. It is always kiln-dried and surfaced four sides. It is used for interior decorative purposes such as shelves, window surrounds, splash boards, cabinet work, and certain door jambs. It is used sometimes for exterior purposes such as fascia and window trim. And sometimes it is run to a tongue and groove pattern for paneling or siding. However, it is nearly always sold as an S4S product.

The supply of wood of this quality is diminishing each year in North America. The Southeast Asian forest industry has it and we can use it. The selling price of "finish" is steady to rising, and is not volatile. It does not require sophisticated machinery to produce it, but it does require careful grading and handling. The grade allows certain defects not permitted in FAS or "select and better," such as sap stain and certain tiny blemishes. One of the major firms in the industry is importing rough 4/4 meranti into the United States, kiln-drying it and blanking it to 15/16-inch. The shipment is then graded to sort out all the pieces which will make "finish" grade. These are run back through the planer to 3/4-inch x finished width, and packaged and sold to the retail trade as a supplement to softwood finish. Those pieces which do not make "finish" are graded according to the NHLA rules and sold to the conventional hardwood users. I understand that at least one company in the Philippines is processing their lauan in this manner, packaging it in small packages, and shipping it directly to retail stores in the United States. Supposedly they are getting a premium price for this because they are giving the customer exactly what he wants.

I mention "finish" lumber because it is an item possible to make right now in Southeast Asia with only a slight change in the usual operation.

New species

Many Southeast Asians are logging in areas that more resemble a botanical garden than a forest, with dozens of species to the hectare. Our market is diverse and flexible, willing to try something new. Therefore, those who are in a position to do so should investigate the possibilities in offering what are generally called the "minor species." Let me give a few pointers on selling them to our market.

First of all, the costs of introducing new species into the market place are very substantial, and the risks are unknown and potentially large. For this reason, your selling prices, in the beginning, will have to be lower than are the prices for the established species. It is relatively simple for us, as importers, to sell a carload of meranti or keruing. Our customers know what these are. We can make the sale with a telephone call. However, a minor species is usually an unknown item to our customer. We will have to make many, many calls to a wide range of prospects until we find someone who is willing to try it. Then we will have to send him samples and provide technical information. There is much more work involved on our part in making this sale, and we must have a higher profit margin to offset these higher costs.

From our customer's viewpoint, he must be offered an incentive to try this new species. If he is offered a choice between lauan or an unknown species at the same price, he will take lauan every time. This is because he knows lauan, it already fits into his scheme of things, and he understands it. He knows how readily he can sell it. But with a new wood he runs the risk of its being unacceptable to his customers and of being stuck with unsalable merchandise. If our customer is a manufacturer, he will have to spend money to determine whether he can process it satisfactorily in his plants. And if he is making a proprietary line of products, such as furniture, he probably will have to spend extra money on an advertising campaign to introduce the new type of wood. Any time you make a change, you increase the costs of doing business, at least in the beginning. This may be one-time-only cost, but it does have to be amortized. I recall several years ago we were in Singapore buying the usual market specifications of meranti. At the time, the price was about $200 per 1,000 board feet, FOB. We inquired about the possibility of also purchasing some of the minor species, and were told that the price was the same as for meranti. We pointed out that these minor species traditionally sold at 50 percent of the price of meranti, and that a price equal to meranti was too high. However, the sawmillers were adamant that it was to be a meranti price or not at all. The result was that we did not purchase any minor species and those trees are still in the woods.

In addition to a lower initial price, there are two other requirements to consider when attempting to introduce a new species. These are volume and continuity. There must be an adequate volume of wood available in order to make it worthwhile for an importer to spend his time developing a market for this new wood. I would suggest that at least 250,000 board feet per month be available on a sustained basis once the business is established. For any amount less than that the profits to be gained will not pay for the costs of setting up the program.

There must also be a reliable continuity of supply. It is a major decision for a customer to take on a new item, and he needs to know that he can count on repeat business if the project is successful. We once were offered samples of an absolutely beautiful wood that could have sold well in our market. However, when all was said and done, only 20,000 board feet were available every three months, and the project was therefore scrapped.

In developing the lesser known species for the American market, I suggest that emphasis be placed on the white-colored woods. Our market prefers lighter colored woods, particularly those with white or yellow hues. Also, there are many more reddish-colored woods available in the world than there are white woods. A newer species will have difficulties breaking into the market place if it is considered to be "just another reddish-colored wood."

I also suggest working only with one or two customers when introducing a new wood. If the project is successful, the supply of wood will probably be totally used up by these few customers. If the supplier works with a number of customers, and the project is successful, he probably will not be able to give them the volume they need. Thus the customers will be unhappy with the supplier and figure he was just another small-town mill that could talk but not deliver.

Finished goods

I use the term "finished goods" to include all highly manufactured or sophisticated wood products. These include mouldings, millwork, and furniture parts. I think the prospects for Southeast Asians to sell such items in the United States and Canada are excellent.

Unlike some other countries, we do not place a great emphasis on the principle that all the conversion and manufacturing work must be done at home. We are results-oriented. We want the best quality at the cheapest price. If someone else can do it better, we will let him do it and then buy it from him.

The economics of freight and waste disposal work in favor of the Southeast Asian industry. Why haul those shavings and trim ends halfway around the world, and then have to pay to have them disposed of? Similarly, Southeast Asian labor is cheaper than ours, and many of these items are labor intensive. And being closer to the log source, the Southeast Asian miller can be better assured of getting the right logs needed to fill orders.

The quantity of domestic timber of the quality suitable for making long-length mouldings is diminishing. Consequently the market for these items will increasingly look toward Southeast Asia. We are making certain progress in getting people to accept finger-jointed and edge-glued stock, but many markets will not accept such stock and may never do so.

There is a continuing demand for furniture parts, regardless of whether they are run-to-pattern, turned on a lathe, or have end-work done to them. Many are currently being produced in Southeast Asia and much of it is made of ramin. We believe there are other woods that could also be used, as long as they are available in large, commercial-sized quantities.

Let me sum up the North American market in a few statements:

1. Quality is the keynote. The only product that has ever withstood the test of time in any market is a quality product.
2. Establish good manufacturing standards and grading rules and stick to them. This is the best way to gain the confidence of the customer.
3. Whatever you do, do it well. In our marketplace, only a handful of operators are identified as doing a first-rate job. When the market is slow, they get first look at all the orders; and in normal times they have all the business they can handle.

Much of the lumber manufactured in North America is used for construction of industrial buildings and private homes, with framing methods as shown.

Finish lumber, like that shown, is made from high quality raw material and is processed to quite accurate dimensions and with a particularly fine finish.

CHAPTER 3

The Market for Tropical Hardwoods—The European Market

J. L. SPEECHLY
Senior Director, Price, Morgan (Hardwoods) Ltd.
London, England

Ever since primitive man set foot on earth many centuries ago, timber has proved to be his friend, and the demand from Europe for this valuable commodity increases almost every year. To cover this vast subject I have decided to ask four questions:

1. Where are the major hardwood producing areas?
2. What are the main uses?
3. Does the demand for individual species vary from one country to another?
4. What does the future hold for the timber industry?

Let us first examine the major hardwood-producing areas. It is calculated that over 25 percent of the world's land surface is covered by forest, and no less than 67 percent of this timber is made up of hardwoods. Naturally these estimates can only be approximate as there are vast areas which have yet to be explored, such as the Amazon. Nearly all the supply of hardwoods comes from four areas which are Africa, Europe, Asia and the Americas, and Figure 3.1 shows the development of these areas since 1955.

The African countries are still the major suppliers to Europe as they ship both logs and lumber, but there are doubts about the political stability of some of these countries which could undermine future developments.

As the demand for timber in Europe continues to grow it can be reasonably expected that countries such as France and Germany will prefer to consume their own production, but I believe that countries such as Romania and Yugoslavia are certain to continue to export their timber in order to earn foreign currency.

Prior to World War Two, North America was a major supplier of hardwoods to Europe but the situation has changed considerably due to the economics involved. Many people feel that South America will play a very prominent part in the future hardwood business and I go along with this belief. However, I think that it will be at least a decade before this area comes seriously into the picture as its forests contain such a mixture of species. It has been calculated that there are up to 150 different species in many of the stands, which obviously makes it very difficult and costly to market individual species according to the present format of the trade. However, this area will develop quickly as soon as Europe is prepared to accept a group of species according to colour, weight, strength, or other characteristics, and I believe that this sytem is likely to be put into effect within the next 10 to 15 years.

There is little doubt that Asia and the Pacific will supply an ever-increasing proportion of the European hardwood requirements provided there is political stability. Figure 3.1 clearly shows that Africa has supplied consistently about 40 percent of the hardwoods imported into Europe, whereas Asia has gradually improved her position from just under 24 percent in 1955 to 34.8 percent in 1974. While forests in Malaya, the Philippines, Sabah and Sarawak are being extensively worked, there are areas such as Kalimantan, Indonesia, Sumatra, and the numerous Pacific islands which are only just being developed. There is little doubt that continuity of supply from the Far East is guaranteed for many years to come.

End uses of timber

Moving on to the end use of timber in Europe, we soon realise that wood is extremely versatile and suitable for literally thousands of different purposes, but broadly speaking, these can be placed into three distinct categories: construction, utilitarian, and decorative work.

Hardwood that is to be used for structural purposes has to have the necessary strength properties and be competitive in price not only in comparison with softwoods but also against other materials such as concrete and steel. Colour and other factors such as appearance mean very little to users of timbers in this category and there is a tremendous growth market in this field, especially for species such as keruing, kempas, kapur, and meranti.

Hardwoods have continued to hold their place in the market demanding timber for utilitarian purposes in spite of the competition from sheet materials, plastic, and metals. There are no great demands as far as strength is concerned and colour is not important. However, timbers in this category must be capable of taking staining fluids, machine easily, and produce a good paint finish. Ramin is an ideal species for this particular end use.

The decorative species, such as teak, do not need any explanation as it is obvious that their appearance is all-important. It is noticeable that in Europe more and more timber is being used in hotels, offices, and private homes not only due to the price factor but also because people are now beginning to appreciate the fact that natural products will always be more attractive than "man-made material." I believe that gradually it will be understood that timber has its own characteristics which are often enhanced by making full use of what many people call defects, such as knots, wild grain, and a variation in colour. Naturally, timber that fits into this category will always be subject to the changing demands of fashion but surely wood can meet any situation as it can offer so many variations of colour and texture.

Continuity of supply is of tremendous importance, as one can easily imagine the severe losses which would be incurred if a furniture factory or building site was brought to a halt due to a breakdown in the delivery of timber. This is where our trade will continue to benefit as species from each of the major producing areas can so easily be interchanged, depending upon the political and economical climate which exists in any one country at any particular point of time. Let us take a few examples. If one is looking for an excellent joinery timber, meranti is available from

Asia, utile from Africa, beech from Europe and mahogany from South America. Should we require a good utilitarian wood, ramin serves our purpose from Asia, abura or obeche from Africa, beech from Europe and virola from South America.

In Figure 3.2 I have endeavoured to place the end use of timber into six different groupings, and I have included the decorative timbers in various sections owing to the fact that they are used in most of these groups.

Quite obviously the construction and furniture industry takes over 80 percent of all the hardwood consumed and I would like to discuss these various groups.

The building and civil engineering trades need constructional timbers for exterior and interior joinery which include window and door frames, window sills, cladding and a multitude of indoor uses such as staircases and skirting. Perhaps timber will also become more popular for flooring as buyers realise that carpets are so much more expensive. Hardwoods are also required for repair and maintenance work and certain specialised species are needed for marine and fresh water construction.

At one time the railways were major buyers of hardwoods, but plastics and sheet materials have made considerable inroads into this market. However, the advent of containers has created a new demand for timber, and species such as keruing have proved ideal for this purpose.

Shipbuilding used to consume a considerable amount of hardwood, but demand from this sector has now waned. The recent popularity of sailing and motor boats could improve demand for the more decorative species in due course.

With the cost of labour escalating in Europe, many homeowners are now doing their own simple joinery work and consequently the demand for dowels and mouldings has increased considerably. This business is termed "the do-it-yourself trade" and ramin has proved to be once again ideal for this purpose. The majority of kitchen units are made of this species due to its clean appearance.

Among other uses for hardwoods are the manufacture of crates and pallets, pattern making, and the production of sports equipment, toys, and games, but as I said earlier, the use of timber is almost limitless.

Lumber specifications and import preferences in Europe

I now wish to say a word or two about the specifications required in Europe. Although wide variations of thicknesses are imported, 1-inch and 2-inch are by far the most popular, especially in the United Kingdom. The thinner sizes are sold to manufacturers in the furniture industry, whereas the thicker stock is required for joinery work and construction. There is an increasing demand for fixed widths and lengths and already 48 percent of all lumber imported into Europe is in fixed widths and 17 percent in fixed lengths. In addition, about 15 percent is now imported in small dimension sizes. The demand for fixed specifications varies from one species to another and it is interesting to note that over 80 percent of all the keruing imported into Europe is in fixed widths, 68 percent of ramin, and 48 percent of meranti.

The law of supply and demand also plays a great part in the sizes used and the prices paid. An excellent example is 1-inch by 5-inch ramin. At one time this size was included in a general specification. However, a very large kitchen furniture maker in the United Kingdom found that this was an ideal size for its requirements and decided to buy only this one width. Within a few weeks the price for this item had increased by at least 50 percent in comparison with the general specification. However, the company recently decided to cut back its purchases and almost overnight the price for 1x5 ramin fell back to its former level.

At one time, stair treads were 2-inch by 9-inch or 2x12. When prices increased, builders found they could use 1¾x9 or 1¾x12 and I now understand even 1½-inch thickness is being used. Another example is window frames in Holland. The normal size used for this purpose was 3x5. Resulting from firmer prices, this section was reduced to 3x4½, then 3x4, and now many builders use 2½x4.

Although only about 7 percent of the hardwoods brought into Europe are machined prior to shipment, an appreciable quantity is moulded and planed by the importer before it is delivered to the eventual user. With the advent of containerization and the fact that buyers are very much aware that it is preferable to leave the waste material in the country of origin, thus reducing the cost of freight, it is not difficult to forecast that this section of the trade will develop considerably over the next few years. The only possible inhibiting factor is that many factories have their own moulding and planing machines and will be loathe to dispense with this machinery until they are confident that overseas shippers can give them continuity of supply. The unions are extremely strong in Europe and many end users have encountered difficulty where they have decided to reduce the output of their moulding mills, thus causing unemployment. It does appear that many factories have partly overcome this problem by ordering some of their requirements from abroad while continuing to produce the major portion of their needs from their own machines.

As far as the demand for kiln-dried timber is concerned, only 5 percent is imported in Europe in this condition, although over 35 percent is kiln-dried prior to delivery. There is no doubt that the cheaper labour which is available in many of the producing countries will soon bring about a greater need for timber which has been kiln-dried at source, and I think that Asia has a great advantage in this respect, as there is at present very little kiln space available in Africa and South America, and the kilns in Europe are very expensive to operate and usually in great demand.

Although I have given the overall picture of the uses of hardwoods in Europe, tradition makes some species much more popular than others in individual countries. Figure 3.3, which shows these differences in popularity, is based on the exports from Malaysia and Singapore for 1973, as at the present time all the figures for 1974 are not available.

The reason why merbau is so popular in Holland is very interesting, and in fact this country imported 82 percent of the total shipment of this species which was 41,000 tons. The Dutch have been involved in Indonesia for many years and it is only recently that they have moved out of the area. As a result of their presence, Java teak was in great demand for making doors in Holland, but recent political developments and the escalating price of teak forced the Dutch builders to look for an alternative species. They found this in afzelia from West Africa, but the quantity was not large enough to cover demand. It so happens that merbau has the same family name—*leguminosae*—as afzelia, and therefore this species soon became popular for doors.

Naturally builders started to use it in place of teak for other joinery purposes, and gradually merbau was used for making stairs and eventually for window frames and other joinery work.

By far the largest demand for keruing comes from the United Kingdom who took about 67 percent of the total 128,000 tons exported to Europe. Britain uses this species for many outside purposes such as agricultural buildings, but there is an increasing demand for keruing to make trucks, trailers, and the flooring for containers. The most popular sizes are 1x6, and 1¼x6 in lengths of 16 to 18 feet. At one time the United Kingdom imported large quantities of keruing to make the framework of railway wagons but unfortunately this market has now been lost to steel.

France took over 50 percent of all the kapur brought into Europe which totaled 44,000 tons and 63 percent of the mengkulang totaling 18,000 tons. While many countries use tropical hardwoods for show purposes, most French builders prefer to paint it. As you know kapur is prone to wormholes, and in many cases is sold on the basis of "pinhole no defect." However, it is possible that the French have overcome this problem, as it is a comparatively simple matter to fill in the holes before painting.

The most popular species in Europe is dark-red meranti and over 502,000 tons was imported in 1973. Holland took just over 36 percent, France 24 percent, and Belgium 16 percent of this total. Most of this species is used for joinery and its popularity is very closely linked with the price of utile from West Africa, which is also an excellent joinery timber.

In the United Kingdom dark-red meranti is used on a very much smaller scale due to the fact that most window frames are made out of softwood or metal, and also because utile is preferred for the better joinery work. However, Britain used over 64 percent of the total quantity of light-red meranti which amounted to 41,000 tons in 1973. This timber is used extensively in the interior joinery trade, for such purposes as door frames, skirting, and many different types of mouldings.

Italy used 73 percent of the total of 25,000 tons of jelutong which was exported in 1973, and the United Kingdom took 22 percent, but these two countries use this species for very different purposes. The furniture industry in Italy requires light-coloured timber and consequently large quantities of 1-inch, 1¼-inch and 1½-inch ramin is imported for this purpose. Beyond this thickness ramin has a tendency to become stained in spite of dipping, and therefore jelutong is used in the thicker sizes. In the United Kingdom yellow pine from Canada is ideal for making detailed patterns prior to the actual production of machinery as it is so very stable. However, this species is extremely expensive and jelutong is used where cheaper timber can be utilised as it also has the same properties.

Out of a total of 266,000 tons of graded ramin exported to Europe in 1973 Britain took 35 percent, Italy 30 percent, and West Germany 23 percent. These countries all use this species for furniture and mouldings. West Germany also uses large quantities for door frames as the Germans prefer to have light-coloured doors. This explains the considerable demand for 7 and/or 14-foot lengths from this country.

At one time Finnish birch was used almost exclusively for furniture frames, which are eventually covered with fabric, but due to the shortage of supply and a rapid increase in price, British buyers turned to sepetir which has proved very adequate for this purpose. In fact the United Kingdom imported 88 percent of this species out of a total of 11,000 tons. This is yet another example of the interchangeability of timber.

Looking at the other European countries one wonders why Portugal and Spain do not import sizeable quantities from Asia. In my opinion the answer lies in the fact that these countries, until recently, were able to get their timber requirements from their colonies. In view of the latest developments I wonder whether they will not turn to the Far East for their requirements or to South America. I believe there must be a good growth market potential in these countries for tropical hardwoods.

It appears that the eastern European countries import very little or no timber from the Far East and one can assume that it is unlikely that the position will change.

I would like to emphasise that while the figures I have given you illustrate quite clearly the individual demands from the various countries, there could be a certain amount of distortion. For example, timber imported through Amsterdam and Rotterdam could be destined for inland users in Germany. Nevertheless, I think that the figures illustrate very well the points I have been endeavouring to make.

The future for timber

The future for timber is very exciting. It has often been said that the ever-increasing demand for timber will eventually cause the prices to rise to such an extent that customers will be forced to switch to substitute materials, such as metals, concrete, and plastics, but recent developments have been very favourable for the timber trade.

When comparing the cost of producing wood with the cost of manufacturing alternative materials, we must consider the fact that the energy requirements for producing substitutes are between three and eight times as high as for the production of timber for similar end uses. This factor alone makes timber very competitive while the cost of fuel remains so high. Furthermore, in most cases, substitutes for timber cannot be renewed whereas it is a fairly familiar sight for one to see demolition contractors reclaiming joists and spars from old buildings which have been demolished.

Our society is becoming more and more conscious of pollution and this again turns in favour of timber, as it cannot be denied that air, water, and land pollution, resulting from the production of substitute materials, is much higher than that arising from the production of wood products. The level of pollution can obviously be brought down considerably, but only at high cost, which inevitably benefits timber.

As mentioned earlier in this chapter, I believe that the grouping of timber into its suitability for specific purposes will develop in the near future, and I feel that stress-grading will play an important part in this process. Already the Timber Trade Association in the United Kingdom is carrying out extensive tests along these lines. One can easily imagine the advantages timber would have over other materials, when the consumer no longer has to decide on the individual species suitable for his requirements, but can order a certain grouping which he is confident will provide the right wood for his purpose, regardless of whether it has come from Africa, South America, Europe or the Far East.

I believe that the present trend to make fuller use of forest products will continue, but it is absolutely ludicrous that a great deal of time and money is spent on building roads, railways, and developing communications into the jungles of the world in order to select a limited number of species. Surely the time will come when a whole stand of timber will have to be harvested and the area cleared and replanted. Bearing in mind the fact that it is not always possible to construct timber complexes near the sources of supply, due to lack of suitable labour, shortage of finance, or the difficult terrain, highly developed areas such as Singapore and Hong Kong could play a very important role in providing processed timber as they are ideally suited to supply the technology, finance, and labour required.

At the present time Africa and Indonesia are supplying the major portion of all logs imported into Europe, but I forecast that supplies will soon be reduced, as it is obvious that it is more financially rewarding to process the logs in the countries of origin, as more labour is utilised and a higher income obtained. However, there will always be a certain demand for logs for special requirements such as producing articles that must have timber of identical colour and texture. In cases such as this I imagine the buyers will wish to inspect the logs at source and will probably pay very good prices. We are already witnessing this process in the veneer trade.

Availability of freight is of paramount importance to the timber-producing countries. Africa is in the most favourable position as far as the cost of freight is concerned, as the lumber freight rates from there to Europe average about U.S. $61 per cubic metre, whereas rates from South America average about U.S. $65 per cubic metre, and from Southeast Asia about U.S. $70 per cubic metre. Owing to the very much longer distances involved in carrying timber from the Far East, these differentials will always remain, but they could be more than compensated by the higher level of development existing in Asia, which places the area in a favourable position to take full advantage of containerization. To obtain the best possible freight rates, I feel sure that the timber trade will have to adjust to the requirements of the shipping lines as these companies endeavour to bring in regulations to ensure the speediest possible turnaround of their vessels. We have already seen this development in the form of containers and stricter bundling requirements, and I am sure that even more efficient methods will be found to carry hardwood to Europe. In return for the timber trade adjusting to different handling methods, I am confident that the shipping companies will provide regular freight space for the large volume of timber that will be required by Europe.

The successful modern timber operation involves forest surveying, building of roads, logging, the construction of kilns, moulding mills, and veneer and sheet material plants, and there is no doubt that a vast amount of finance is required. Obviously, developing countries will look for assistance from outside sources to provide the necessary finance to make full use of their natural wealth. I believe there are many interested investors but right conditions must prevail for this investment to take place, and I am rather disturbed that the political instability that exists in some of the hardwood producing areas could restrict the further development of these regions.

The day will surely come when it will be necessary to work out a grand reforestation programme on an international scale, but I doubt whether the authorities concerned will treat the matter with urgency, while the statisticians continue to assure us that more hardwoods are growing than being consumed.

In my opinion, the demand for timber throughout the world will continue to increase, especially as underdeveloped areas become more prosperous, resulting in the people demanding better housing conditions and subsequently more furniture, both major timber users. Therefore, ladies and gentlemen, I consider that we are all very fortunate to be concerned with an industry which not only allows us to handle a natural and sometimes beautiful commodity, but also to be involved in a business which offers almost limitless opportunities for development.

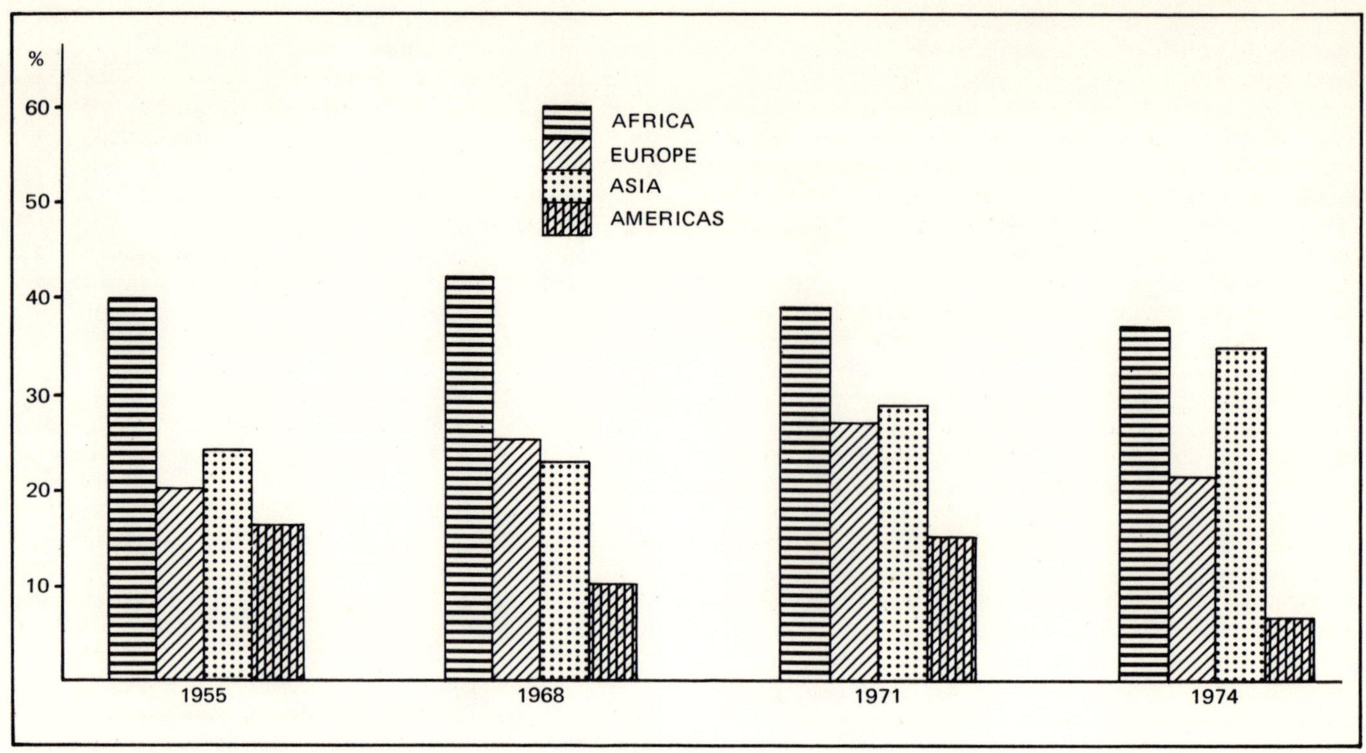

Figure 3.1. Percentage distribution of European hardwood imports by area of origin.

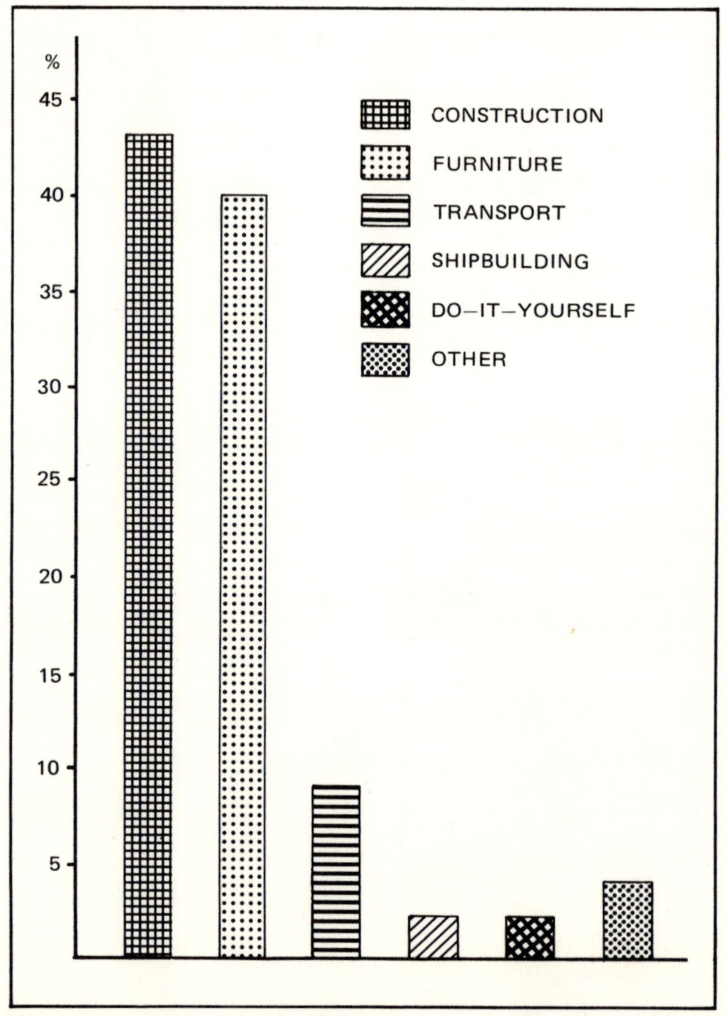

Figure 3.2. End-use of tropical hardwoods in Europe.

SAWMILL TECHNIQUES FOR SOUTHEAST ASIA

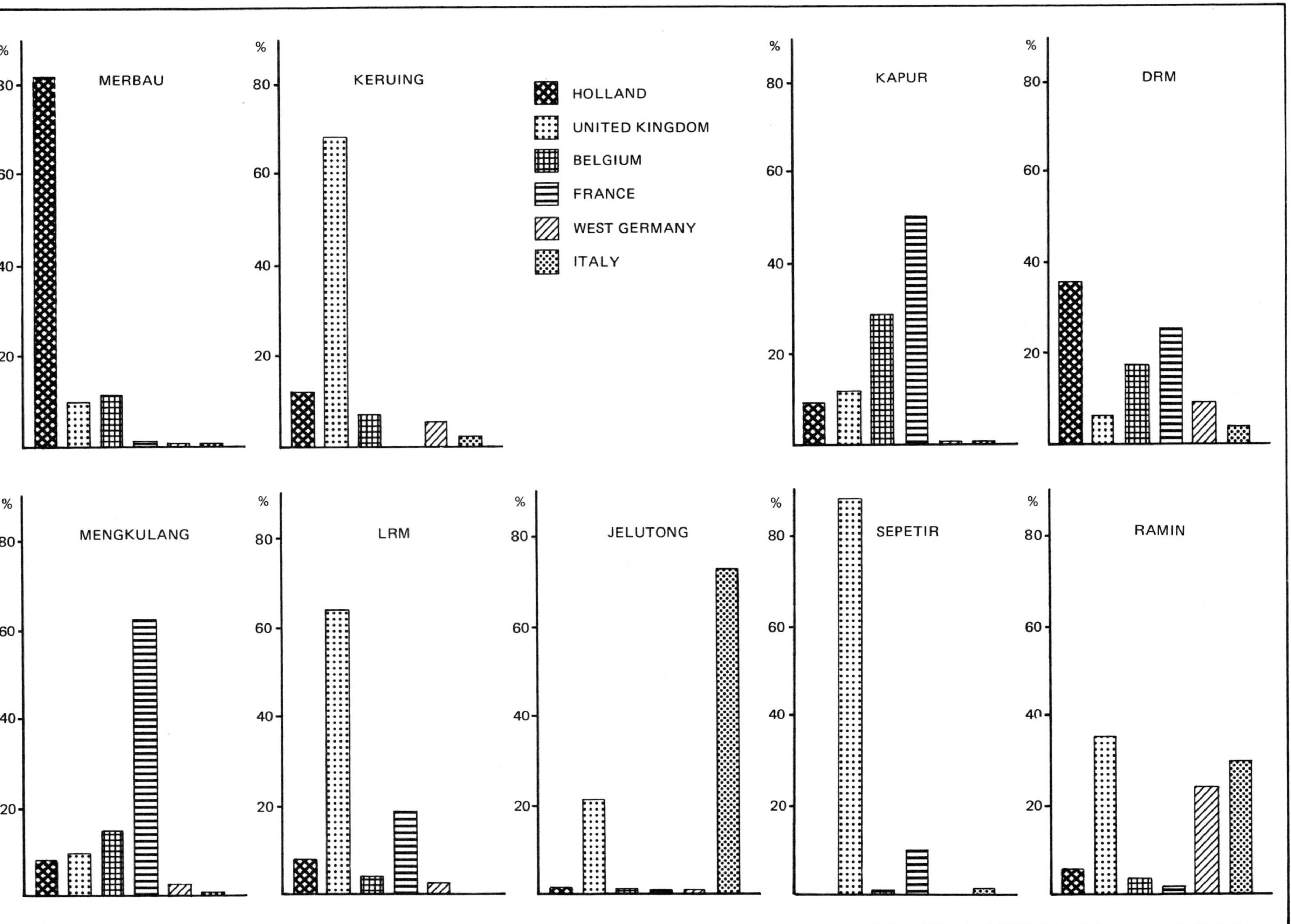

Figure 3.3. Percentage distribution of major species to European markets.

CHAPTER 4

The Market for Tropical Hardwoods—The Japanese Market

MIKIO "MIKE" SASAKI
Managing Director, Sasaki Co. Ltd.
Tokyo, Japan

While most persons in the wood products industry realize the enormity of the wood products market in North America, fewer realize the potential of Japan as an end user of wood products.

As an example, in 1973 the total number of houses constructed in Japan was 1.8 million. Japan's future needs for housing will be large.

Effects of the "oil-shock"

Until two years ago, the Japanese economy had been expanding at a high rate. This high-rate growth was considerably slowed down by the so-called "oil-shock"—first, the shortage of oil, and then the high cost of oil. Last year, housing construction in Japan dropped. However, because Japan's latent demand for housing in general is so strong that, when her economic situation turns brighter—we hope in the near future—the housing industry will be among the first to improve. After six months of the "oil-shock," the prices of almost all construction materials have increased to twice as much as before. Labour costs are also skyrocketing. The Japanese government has placed a strict control on housing loans available from banks. All these factors have made it very difficult for the average Japanese to realize his long-time dream of having his own home before retirement.

Housing conditions in Japan

The total number of houses built in 1973 in Japan was 1.8 million, while in the United States 1.9 million houses were built in 1972. This means that the number of houses constructed in Japan in the 1972-1973 period per 1,000 people was two times larger than that of the United States.

It is estimated there are about 12 million families who do not own their own homes. Of the 12 million families, 9 million are said to want to have their own houses built.

The second largest group are those who already own houses but are dissatisfied with the small space and poor facilities. These are the people who want to have larger houses rebuilt on the same site equipped with air-conditioning and other modern facilities. It is estimated that six million families belong to this second group. With the two groups put together, we have 15 million families, or 50 percent of all Japanese families, who want to have houses built or rebuilt as soon as they can afford to do so.

Many young couples and their families are living in apartments provided by a government agency, the Japan Housing Corporation. Most of the public apartment buildings are of five stories and are made of reinforced concrete (see Figure 4.1). Older apartments are called "2DK's," which means two rooms plus one dining room-kitchen

combination. "D" stands for dining room, and "K" for kitchen. Newer ones are a bit more spacious, they are "3DK's," meaning three rooms plus one dining room-kitchen combination.

Although Japanese people are generally smaller in size than Americans, they are not satisfied with the small space of the public apartment which is only 550 square feet altogether. Many of them want to move into a house of at least 1,000 square feet, and preferably with a garden, however small it may be. To these people, having their own home is one of the important goals in their lives.

Small private houses, such as shown in Figure 4.2, are made of wood. Many of them are prefabricated. One of these tiny houses would cost about $20,000. The tiny piece of land on which they build is described by Japanese as "no larger than the forehead of a cat," and would cost about $15,000. These are typical houses seen in the outskirts of big cities like Tokyo, Osaka, Nagoya. The occupants are usually those in the middle-income class who have to suffer through a two-hour train ride to get to their downtown offices. At any rate, these are the only houses that are within the reach of the average Japanese white-collar and blue-collar workers.

The large house in Figure 4.3 would cost about $70,000. The land, which may be 1,000 square feet in size, would cost about $50,000. Only such high-income earners as large corporation executives are able to buy such luxury.

The "2x4 system"

Last fall, the Japanese Ministry of Construction finally approved that any contractor can use the 2x4 framing construction system. In Japan we call this new system simply the "2x4 system." It is estimated that by the end of 1975 3,000 houses will be built by using this new system. Actually, maybe only half this number will be built, because of the shortage in the number of carpenters skilled in the use of the 2x4 system.

Therefore, the number of houses built on the 2x4 system is very small in comparison with 1.8 million houses being built yearly on the conventional system. Within five to ten years, the 2x4 system will probably become a major system in Japan for building private houses.

The reason is simple: *labor.* Let us consider the present labor situation in Japan in the field of construction. There are two factors: quality of labor and quantity of labor.

It takes seven years for a young Japanese boy to become a professional carpenter able to fully use those sophisticated techniques of the traditional "post and beam" system. However, young boys are no longer interested in construction business. More than 95 percent of junior high school graduates, who are 15 years old, go on to senior high school. Most of the remaining 5 percent, half of them girls, are unqualified to do the senior high school studies. If they

---

This paper was delivered at the Seminar by Duane C. Mason, Associate Editor, *World Wood.*

cannot do the three-year senior high school studies, how can we expect them to complete the seven-year special training to become a carpenter? Therefore, we cannot expect many carpenters-to-be from this group.

They could start carpenter training after senior high school, but almost all the graduates are interested in the so-called "high-collar" careers. Nearly 40 percent of the senior high school graduates go on to college and the number of students in this classification is increasing rapidly. Therefore, we cannot expect to find many carpenters-to-be in this group.

Thus, the future of construction labor in Japan is very dark. And this is the very reason why the 2x4 system attracts so much attention of those concerned with construction business in Japan.

Figure 4.4 illustrates Japan's conventional house-building system but using imported 2x4 U.S. lumber. A housing developer in Tokyo has built 1,000 houses in this method, probably because of the shortage of lumber during the 1972-74 period in Japan.

Japanese distribution system

Now I would like to tell you something about the Japanese distribution system. It is complicated. Before lumber reaches the individual contractor, it has to go through six to ten wholesale agents, and at each wholesale agent the price of lumber is increased by 10 to 20 percent. This is true with imported lumber. Unless the whole distribution system is reorganized, the cost of houses will never go down.

Canada has been active in the sales promotion of their 2x4 lumber. To show the public how their 2x4 lumber can be used, the Canadians built a residence last fall attached to their embassy in the heart of Tokyo. It is said that they are planning a free workshop for Japanese carpenters in which they will demonstrate how to use 2x4 lumber.

You will find the Japanese people in general are fond of houses made of wood, are fond of the odor of wood and of the grain of wood. I think you will find an endless need in Japan for good quality lumber.

Figure 4.1. Most public apartment buildings in Japan are five stories made of reinforced concrete.

Figure 4.2. Small homes in Japan are made of wood.

Figure 4.3. This house would cost about $70,000 in Japan, and the land would cost about $50,000.

Figure 4.4. This house is being built using the conventional Japanese system, but with 2x4 lumber imported from the United States.

CHAPTER 5

Grading Hardwood Timber Reliably

EDWIN B. HUDDLESTON
Consulting Engineer
Sydney, Australia

Timber comes from logs. Logs come from trees. Not every tree produces a saw log and in those which do, it is likely that some lengths cut from the tree will not be suitable for saw logs. In every saw log it is necessary to turn it and selectively cut it to produce the timber required from it. The volume of sawn timber which can be produced from the log is very largely controlled by the number, location, and size of the imperfections acceptable in the sawn timber. Furthermore, the amount of work required to obtain the timber from the log is greatly increased as the limitation on defects permitted in the timber becomes more restrictive.

In the case where defect-free material is required, the total volume obtainable from the saw log may be as low as 30% or even less of the total volume of wood in the saw log. The same log sawn to a reasonable specification may produce in sawn timber as much as 80% of the timber in the log.

Recovery from the log greatly influences the cost of the sawn timber. If it costs a figure of X dollars to buy the tree, to put in roads and snig tracks, to cut and remove the log to the sawmill, and to saw the log in the sawmill, the cost of the sawn timber produced at 30% recovery is 3 1/3 X dollars and in the case of the 80% recovery is 1 1/4 X dollars. The difference is added to by the extra cost of sawing to the more restricted specification. It is very large and compels that imperfections be permitted.

To allow defects in the sawn timber requires some system to ensure that unacceptable imperfections shall not be included in the material delivered to the buyer. This system is called grading. Some grading is accomplished by production personnel, but basically the grading system is one which subjects the timber to an examination and classification after it leaves the mill.

The basis of a timber grading system

Any satisfactory timber grading system will have three essential components. They are:

1. Satisfactory specifications covering the material to be graded.
2. Competent timber graders to inspect each and every piece of material or article submitted for grading and to determine whether the particular item meets the requirement of the specification.
3. A check inspection, or quality control system.

The specification is a compromise between the buyer, who desires to obtain the best available material at the lowest possible price, and the seller, who hopes to sell as much of the log which he uses at the highest price he can obtain for it.

These conflicting interests lead to an agreement which becomes the specification, and this in turn becomes the basis of the contract of sale. Normally the timber grader has no knowledge of the decisions made or the reasons which caused them to be made in preparing the specification. It is not part of his job to query those decisions, which come to him as part of the specification, and in fact he has not the capability to do so.

The specification given to the timber grader may be completely different from one to which he is accustomed for a similar product. Again it is not part of the grader's job to determine whether one or the other of the specifications is correct. His task is solely to ensure that the provisions of the relevant specifications are met by the piece of timber or the article which he grades as complying with it.

Should the grader find that he cannot understand or interpret any provisions of the specification, it is his duty to seek directions from the person who gave him his instructions to carry out the grading, and from no other person. Normally such interpretation should be in writing, by way of amendment to the specification, so that any person called upon to carry out a check inspection will be made aware of it.

A check inspection, on a sampling basis, is necessary to maintain the skill and competence of the graders. Normally we would expect that our graders are competent and that they seek to conscientiously carry out their work. However, all parties to the contract have a vested interest to ensure that any grader who may succumb to inducements, or who through laziness or other cause, does not grade timber in accordance with the specification is quickly brought into line or removed from his job as a grader.

This is only one reason, and a minor one, for check inspections. More important, it is very difficult for a competent grader to maintain a uniform standard. He must continually make quick decisions as the timber is presented to him and the quality of the parcel as a whole tends to greatly influence his judgement without his realising it. He needs some system of quality control to bring him back to the proper interpretation of the specification should he depart from it, and any good grader should welcome such a system.

Timber specifications

The specification is the foundation of any satisfactory timber grading system. On it rests the success or failure of the timber grading system. Specifications should be written in a clear and concise form, but their language should be kept as simple as possible. Timber graders and production personnel usually do not possess the educational advantages of the business executive or technical personnel who are mainly responsible for the preparation of the specification.

As well as keeping each specification clear, concise, and simple in language, it is extremely desirable that some uniformity be maintained in format, definition of terms, and methods of describing and measuring defects between the various specifications which individual graders or inspectors will have to use.

Definitions and methods of measuring defects can be given in each of the individual specifications which might be written, but this practice encourages slight differences between individual documents and these add to the difficulties of the timber grader.

The grader also finds it most disconcerting to find two different specifications for similar products; for example, sawn boards for dressing and sawn structural timber, using different methods of description, or where the same method of description is used, the relevant clauses being placed in a different order.

It must always be remembered that the timber grader must work at a pace dictated by production processes, and while he must always insist that he is in control of his grading operation, his management would not be happy if he holds up production to refer to his specification. He therefore requires to know exactly where to find the point he needs to check on in his specification, and having found it, to be able to quickly interpret what he finds written.

These requirements are best met by the preparation of some basic specifications. Essential ones are: Terms and Definitions Used in Timber Grading Rules; Methods of Measuring Defects in Timber Grading; Methods of Determining Moisture Content; and Test Methods Used in Timber Grading. Others, such as a specification for surface roughness, may also be needed.

The first of those listed, Terms and Definitions Used in Timber Grading Rules, would seek to list every term which might be used in writing timber grading rules or specifications, and would give a definition of every one listed. Terms such as "spaced," "scattered," and other similar expressions used in specification writing when it is found impracticable to be precise in regard to the occurrence of certain defects, such as borer holes, should be among those listed, and as clear a definition as possible should be given for them.

With the basic specifications provided, all subsequent documents can call them up as part of the document itself, either in their entirety or by reference to specific parts of them. Specifications for products can be conveniently classified as follows:

1. Rough-sawn timber.
   a. for appearance grades.
   b. for structural grades.
2. Dressed products.
3. Manufactured items.
4. Structural timbers.
5. Codes of practice.

The timber grader will be mainly interested in the rough-sawn timber, but certainly in some cases will have to deal with dressed products such as flooring, timber moldings, lining or dressed all-round timber. He will have no function in connection with codes of practice except to the extent that the codes call up graded timber, and in this event his interest will be with the timber specification rather than the code.

He may be called upon to grade manufactured articles or structural timbers. Apart from manufacturing requirements, such as fit and tolerance, these will largely rely on the rough-sawn timber, or dressed timber specifications to define their quality. However, in a manufactured article where the location of the defect can be closely defined, it is often possible to permit in a specified location a defect larger than that permitted in the specification for the rough-sawn timber out of which the article is to be made. In some cases, specifications for rough-sawn timber in manufacturing grades may be prepared.

The main task of the timber grader will be with the rough-sawn timber, and dressed timber specifications, and it is proposed to confine this discussion to the rough sawn timber specifications.

As already indicated, these divide themselves into two groups, the first of which concerns itself mainly with features which affect the appearance of the timber and the second is designed to provide timber of a designated strength ratio. In each group considerations other than appearance in the one case and strength in the other case must be taken into account in writing the specification. Hence, even in the case of the structural timbers where the strength ratio selected for the particular grade will largely determine the size, position and frequency of any particular defect, considerations of fixing, processing or handling may lead to a maximum size for a particular defect smaller than that which would be permitted if the choice was dictated by strength alone. Furthermore, in every case some defects which do not have any marked influence on strength will be permitted and restricted in size, frequency or location in the piece according to the judgement of the parties preparing the specification.

In the case of specifications principally concerned with appearance, the scope for agreement between the parties is far greater and this can lead to two or more different specifications apparently covering the same product. Considerations of strength and fixing in this case may cause a prohibition or limitation to be placed on a defect which from considerations of its effect on appearance of the piece would have been acceptable.

When writing the specification, the aim should be to list all prohibited defects and to give for each type of acceptable defect whatever limitations as to size, position in the piece, and frequency are considered to be acceptable.

With difficulty in some cases, it is possible to define the maximum size, the position, and maximum frequency of individual defects. It is even possible to lay down the distance which will separate defects of maximum permissible size. When, however, different defects occur in combination, each of less than the maximum permitted size, it is impracticable to be precise. The specification must rely on the grader's experience and expertise to assess whether the combination is equivalent or less than the effect of a single defect of maximum permissible size. Suitable clauses to this effect are usually included. As far as possible, leaving such decisions to the grader should be avoided. Where a grader is required to exercise his discretion or to form a judgement, as much guidance as possible should be provided to him by the specification.

Timber graders

As previously stated, the function of the timber grader is to read and understand the specification given to him. Having done so, he is required to examine each piece of timber submitted to him for grading and to decide whether it meets the requirements as laid down in the specification to which he is grading. He must also accept the responsibility of either himself marking the accepted timber with a brand to indicate the grade to which he has assigned it, or of supervising and accepting responsibility for marking.

In fairness to the grader himself, and certainly to facilitate the effective operation of the grading scheme, the grade mark should be associated with some form of identification so that the grader responsible can be determined for as long as the brand remains on the timber.

The grader must accept responsibility for his grading. He must therefore be in complete control of that part of the grading process for which he is responsible. If the timber is coming to him at a rate which is too fast for him to effectively grade it, he must be empowered to slow down the rate at which it comes to him. If the timber is presented to him in a manner which prevents him from seeing all sides of it as effectively as he considers necessary, he must cease grading until the faults are rectified.

The employer has a part to play in effective grading. The grader may slow down a grading chain, or require more lights, or have some other requirements. Such demands are all too frequently met with a threat by the employer to engage someone else if the grader cannot function under the conditions provided. This places the grader in an extremely difficult position, and is all too often the cause of inefficient grading. Some system of licensing graders and the withdrawal of the license for unsatisfactory work is helpful, particularly if the supply of graders is not greatly in excess of requirements.

Selection of personnel for timber grading

The selection of people to undertake timber grading is a very difficult task. In the first instance it must be realised that the grader is part of the production process, in the same way as the sawyer or any other process worker. Care must be taken to select a man with the same sort of educational and cultural background as the other workers in the production line.

On the other hand, we are looking at a man who, because of the work he does, can have a very marked effect on the profitability of the enterprise, and who is very susceptible to the offer of inducements to influence his work in favour of one party or the other. We require a man with the ability to read, understand, and remember the provisions of sometimes complex specifications and to then apply those provisions to the conditions which he observes in a piece of timber. We require a man who has sufficient experience to be able to quickly observe indications of defects in the timber and from those indications to assess the nature and magnitude of the defect present.

Two methods have been tried to obtain the man required. The first of these involves the selection of a suitable experienced man, say a sawyer, from a sawmill and to subject him to a course of training in the interpretation of specifications, the identification of defects, and the effect of defects on strength.

The second is to select a lad from school with an education somewhat superior to that of the ordinary mill worker and to put him through a planned course of training embracing wood technology, personnel management, and timber grading. After completing his academic training, if such it may be called, he is then submitted to practical training in timber grading, finally to be given a grader's

certificate and turned loose on the sawmilling industry.

As might be judged, I am all in favour of selecting the experienced sawmill worker and turning him into a timber grader. If a man has reasonable intelligence, the comprehension of timber specifications can be taught to him. Some associated instruction in elementary wood technology will provide him with the necessary base on which he can build.

The major difficulty is to teach him to be able to work with the accuracy and speed which is necessary in timber grading. This is best done, after his initial training, by using him as a grader with and ahead of another experienced grader. In this way he grades timber, and any which he misses and any mistakes which he makes are picked up by the man following him. As he acquires efficiency, he should progressively be transferred to grading under supervision and then to grading on his own, initially with more frequent check inspections than would normally be applied to an experienced grader.

Check inspection and follow up with grader

The need for a system of check inspection and the reasons for it have already been given. The check inspection on a percentage basis should be carried out by an independent body which, ideally, should share some of the responsibility for the grading.

There are such organisations in various parts of the world which register trade marks and license operators to use those marks. Conditions are applied to the use of the registered mark, such as the employment of qualified inspectors, the identification of the grader responsible by brand on the timber, and an indemnity given to the owner of the registered mark. In turn, the owner of the mark gives a guarantee that the timber carrying that mark is up to the grade marked upon it. He usually covers such a guarantee by insurance against any claims resulting from the guarantee.

Ideally the owner of the registered mark should be able to certify graders and, in turn, withdraw the certificate for faulty grading. Each check inspection should be covered by a report, and a copy of that report should be made available to the grader responsible for the grading being checked. Where the check inspection reveals serious departure from the specification by the grader, the grader should be called upon to show cause why his certificate should not be cancelled.

It goes without saying that check inspections should cover the grader's rejects as well as his branded timber. If this were not so there would be a very strong tendency for the grader to safeguard himself at the cost of his employer by unnecessarily rejecting any piece of timber about which he is in doubt.

The grading machine as a quality control instrument

In the case of structural timber, where the strength ratio for which the specification is written leads to a stress grade for any particular species, or group of species, a timber grading machine has been developed and is finding increasing application.

While its capacity and cost are such that it is unlikely to displace visual grading in the smaller mills, it is possible that, for structural timber, the machine can be used as a quality control instrument and thus avoid the second quality control inspection. By using it in this way it could serve on the timber grader's ability to work to specifications, and possibly the percentage of the appearance grade timber submitted for check inspection could be reduced.

The system visualised is as yet nowhere in operation, but it could well receive earnest consideration if a new grading system is to be set up.

SECTION 2.   GRADE DESCRIPTIONS

2.1 BUILDING GRADE. The timber in building grade shall be truly sawn and free from decay and compression failures, but the following imperfections shall be permitted:

   (i) *Sloping grain*—not exceeding 1 in 8.

   (ii) *Knots.*
      (a) *Sound Knots*—not exceeding one-third of the width of the face of the piece or 4 in diameter, whichever is less.
      (b) *Unsound knots, and knot holes*—not exceeding one-quarter of the width of the face of the piece or 3 in diameter, whichever is less.

  (iii) *Holes.*
      (a) *Not exceeding 1/16 inch diameter*—provided they are well scattered and not clustered.
      (b) *Exceeding 1/16 inch diameter but not exceeding 1/8 inch diameter*—not more than eight holes in any square foot of face of the piece.
      (c) *Exceeding 1/18 inch diameter but not exceeding 1/4 inch diameter*—not more than four holes in any square foot of face of the piece.
      (d) *Exceeding 1/4 inch diameter*—to be measured as knot holes of double diameter (see (ii) (b) above).

  (iv) *Decay in unsound knots.*

   (v) *Sapwood and wane*—not exceeding one-third of the width of face or edge considered (see Clause 1.6.4).

  (vi) *End splits and included phloem intersecting an end*—individual splits or sprands of phloem not exceeding 6 in.

 (vii) *Stain*—if free from decay.

(viii) *Boxed heart*—at the option of the purchaser, but only in pieces exceeding 9 in × 9 in cross-section.

  (ix) *Brittle heart*—not exceeding one-quarter of the cross-section of the piece.

   (x) *Blocked heart shakes*—strands not exceeding 6 in long and spaced not less than 3 ft.

  (xi) *Shakes and end checks*—not exceeding 1/8 in wide or twice the width of the piece long; and not extending from one face to another face.

 (xii) *Surface checks.*

(xiii) *Included phloem not intersecting an end of the piece.*
      (a) *Within 2 ft of the ends*—strands not exceeding 12 in long.
      (b) *Not within 2 ft of the ends, but within the middle half of the depth*—strands not exceeding 24 in long and spaced not less than 1 ft.

Timber grading system specifications should be clear.

SECTION 3.   SCANTLING TIMBER

3.1 BUILDING GRADE. The timber shall be free from brittle heart but the following shall be permitted:

   (i) *Tight gum veins, blotches, streaks and hobnails.*

   (ii) *Loose gum veins up to 1/16 in wide*—not exceeding in aggregate length one-quarter the length of the piece, and no vein extending from one face to another face.

  (iii) *Gum pockets up to ½ in. wide*—not exceeding 12 in. long and not extending from one face to another face.

  (iv) *Holes of any kind, and unsound knots*—
      (a) up to 1/4 in diameter, and scattered—unlimited;
      (b) over 1/4 in diameter but not exceeding one-quarter the width of the face—spaced.

   (v) *Sound and tight knots*—not exceeding in width one-third the width of the face provided knots of maximum permissible size are spaced.

  (vi) *End splits*—not exceeding 6 in long.

 (vii) *Decay and surface termite galleries*—on the surface only, and slight.

(viii) *Slope of grain*—not exceeding 1 in 8.

These specifications are for scantling timber.

SECTION 4.   SCANTLING TIMBER

4.3 BUILDING GRADE. Building grade is intended to provide scantling timber for building purposes where appearance and strength are secondary considerations. The timber shall be generally sound, well sawn, and free from decay, but the following shall be permitted:

   (i) *Sound and tight knots*—not exceeding one-third the width of the face on which they occur.

   (ii) *Unsound knots and knot holes*—not exceeding one-quarter the width of the face on which they occur.

  (iii) *Holes in sound wood*—
      (a) diameter not exceeding 1/4 inch—scattered;
      (b) diameter exceeding 1/4 inch but not exceeding one-quarter the width of the face on which they occur—spaced not less than 3 ft apart.

  (iv) *Tight gum veins, gum spots and gum streaks.*

   (v) *End splits*—one at each end and individually not exceeding 6 in long.

  (vi) *Sloping grain*—not exceeding 1 in 8.

 (vii) *Spring*—not exceeding 1-1/2 in per 12-ft length of piece.

(viii) *Bow*—not exceeding 1-1/2 in per 12-ft length of piece.

  (ix) *Twist*—not exceeding 1/2 in per 12-ft length of piece.

   (x) *Sapwood not susceptible to Lyctus* (see Clause 1.8).

  (xi) *Wane*—not exceeding one-third the sum of the width and thickness of the piece.

  (xii) *Want*—not exceeding one-third the width and/or the thickness of the piece.

Specifications should include definitions and methods of measuring defects in the lumber.

CHAPTER 6

Production Methods for Tropical Hardwoods

MERVYN PAGE
Officer-in-Charge, Forest Conversion Engineering Group
Commonwealth Scientific & Industrial Research Organization
Victoria, Australia

The behaviour of Southeast Asian tropical hardwoods, both as logs during conversion and as timber in use, is generally markedly different from that of the traditional hardwoods of the Northern Hemisphere and of the softwoods of both hemispheres. In this discussion I propose to touch briefly on some of the more important characteristics of tropical hardwoods which affect sawmilling procedures, seasoning methods, or end uses, and to indicate the technology and equipment that has been developed to cope with problems created by these particular characteristics.

Perhaps the most bewildering aspect of tropical hardwood forests is the extremely large number of species that occur, many of them in only small volumes in particular areas, frequently making it difficult to supply marketable volumes of an individual species.

Generally up until the present, loggers in Southeast Asia have been required to harvest only those relatively few species that occur in substantial concentrations and have the characteristics which make them commercially attractive to market and relatively easy to convert. Moreover, usually only those trees of good size and form have been considered for commercial use.

Consequently, relatively little of the forest resource is being utilized, and the forest authorities of Southeast Asia, as well as the more far-sighted timber companies, are actively aware of the need for greater use of both the so-called "minor" or "neglected" species and also the trees of small size and poorer form.

I believe it is likely that the conversion of such material could become obligatory in some Southeast Asian countries in the not-too-distant future, particularly where forest areas are being cleared for agricultural purposes. Large areas of forested land are being clear-felled for agriculture, particularly for the establishment of oil palm plantations. Obviously the community cannot continue to afford to have much of this material simply pushed into rows and burned.

One example of a technological advance which may assist in the marketing of minor species in use groups is called prefinishing. It is a process for staining mouldings and other seasoned and machined items so that they all look the same colour and can be sold for the same end use.

I should also mention that the Food and Agricultural Organization of the United Nations is currently undertaking the listing of minor species from tropical hardwood forests into use groups according to their mechanical, seasoning, and machining characteristics.

Much of the material from minor species and lower quality trees possesses behavioural characteristics which react against the adoption of the high speed production techniques developed for coniferous timbers. In fact some of the trees already being harvested exhibit these troublesome characteristics.

Of course, not all minor species will be immediately utilizable as sawn timber or veneer, but one should not lose sight of the fact that the forests of Indonesia, for example, contain some 4,000 species, while some mills in the tropical regions of Australia are currently processing as many as 80 to 90 species in the one sawmill, and some of these species have very different characteristics.

In general, I believe it is safe to say that in the future the number of species delivered to Southeast Asian mills will increase, while the log quality could decrease. It should be borne in mind, however, that much of the timber produced from these logs may not be offered on export markets but may be used in the construction of low cost timber houses such as the one shown in Figure 6.1. The market for this type of house should markedly increase as living standards improve throughout Southeast Asia. There is already significant evidence that this trend has commenced.

A most obvious characteristic of a tropical hardwood forest is the wide range of log sizes that occur. Some sawmillers may be expected to process log sizes ranging from large reject veneer logs of up to approximately 160 centimeters in diameter down to top logs as small as 25 centimeters top diameter. If the volume in each size class is sufficiently large, then either a special mill or a production line can be designed specifically for that size class. However, the number of very large logs is seldom sufficient to warrant high capital expenditure, and the most effective solution has been to either mill or break down these relatively few logs on an inexpensive portable mill, located either in the bush or in the mill log yard.

The unit shown in Figure 6.2 is a Forestmil. The Forestmil consists of four supporting posts, two crossbeams which are adjustable in height and which in turn support a longitudinal beam carrying a power unit and two circular saws, one vertical and one horizontal. The power unit can be either gasoline or diesel, as desired.

These machines, which with one pass over the log can produce either flitches for further sawing or finished prod-

Figure 6.1. Low-cost timber houses in Indonesia.

ucts, have been very successful in Indonesia, where over 100 are currently operating, cutting a wide range of log sizes. They are being employed in some cases to cut those logs too large foɪ conventional mills, while in other areas they are being used where small, inexpensive, simple and highly portable units are needed to produce flitches for resawing in some instances from logs of marginal quality.

Brittle heart

A fairly common feature of the hardwoods of Southeast Asia is the presence in the centre of the tree of what is called "brittle heart." One form of heart, that which is excessively decayed or eaten away by termites, is readily recognized and is called "pipe." But it is also important to recognize the much less obvious form, that is the area of brittle wood around the decayed heart or, in the absence of decay or pipe, in the centre of the tree. This brittle heart material is often difficult to detect visually, but as it has little or no resistance to impactive or suddenly applied loads, it is unsuitable for many uses and must be excluded from some sawn products. An excellent example of the result of this defect is the tearing-out of the spindle of veneer lathes in some logs of lauan or meranti.

Unfortunately the extent of brittle or decayed wood cannot be readily determined before sawing commences, as the appearances of the ends of the log do not necessarily indicate the occurrence within the log of this particular defect. Therefore this defect imposes two requirements on the type of sawmill that could be considered suitable for converting logs that contain it:

1. It should be possible to dispose of the unwanted material out of the sawmill as cheaply as possible. The sawmill design should provide for such material to be disposed of immediately at the location at which it occurs. It should not be necessary to pass such material from one production machine to another, and to employ the effects of production personnel, simply to get such material out of the sawmill. Figure 6.3 shows one solution, in which a waste conveyor is located directly adjacent to the discharge rollcase of the headrig, but of course there are several other possible solutions, all embodying this general principle.

2. Logs with any significant degree of taper, particularly butt logs, should be taper sawn. In converting a tapering cylinder to pieces which have a regular cross section along their length, it is obvious from

Figure 6.2. Forestmil operating in tropical jungle.

Figure 6.3. Waste conveyor belt located adjacent to band headrig and log carriage.

Figure 6.4, that a proportion of pieces must be produced as shorts. In the figure the amount of taper has purposely been exaggerated to emphasize this aspect.

If sawing is predominantly parallel to the longitudinal axis of the log, then the shorts will be produced in the outside portions of the log; namely that portion which in tropical hardwoods contains the higher quality wood.

On the other hand, if sawing is predominantly parallel to the bark or sapwood, then the shorts will be produced in the centre of the log, that area susceptible to brittle heart.

Studies have indicated that on butt logs taper sawing can increase the yield of first grade boards by as much as 4 percent. Taper sawing requires a carriage with fast acting taper adjustment on the headlocks, permitting the small end of the log to be rapidly moved out towards the saw line and subsequent cuts made parallel to the outside of the log.

Pinholes

Many tropical hardwood trees and logs are susceptible to attack by the Ambrosia beetle, whose attack results in small holes in the wood commonly referred to as "pinholes." For some end uses, pinholes are not regarded as a defect, but for others they are not acceptable. Unfortunately it is not always possible to determine before sawing commences whether or not a log contains pinholes, so that when pinholes are discovered during sawing, the sawmill design should permit sufficient flexibility in production procedures to enable the sawyer to rapidly change from one intended end product to another by, if necessary, rapidly changing flitch size or flitch orientation in relation to growth rings.

Bumps

Another defect which imposes similar requirements on the selection of sawmilling equipment and its layout within the sawmill is the presence of "bumps." Tropical hardwoods are self-pruning, unlike conifers which carry their branches to maturity. The branch stubs that remain from the self-pruning of hardwoods eventually become overgrown and are usually only evident in the mature tree as bumps on the surface of the logs removed from the upper portions of the tree. Sometimes extensive decay can be associated with the overgrowth, but again there is no way of telling before sawing commences whether or not decay is associated with the overgrowth.

Consequently, the sawmilling system employed should be sufficiently flexible to be able to cope with the rapid disposal of low-quality material and allow the sawyer to make an equally rapid change in sawing pattern and, if necessary, intended end product.

Spring

Growing trees contain growth stresses and in some hardwood trees the growth stress level can be sufficiently high and the distribution of stress throughout the trunk such that the end of the log splits on being crosscut and the log and flitches bend longitudinally when ripsawn. This longitudinal bending is called "spring." It is caused by the outside portions of the log being in longitudinal tension and the central or core part of the log being in longitudinal compression. When such a log or flitch is separated by a saw cut, the two portions bend, the bend always being towards the bark or sapwood of the log.

Imagine the outside portion of the log being like stretched elastic, while the inside or core portion is like compressed springs and the forces in the two portions are balancing each other so that the tree or log remains stable. When this stability is upset by a sawcut, the outside portion of the log, that part like stretched elastic, tends to shorten when released, while the core portion of the log, that part like compressed springs, attempts to increase in length. The ultimate result must be a longitudinal bending towards the outside of the log.

If you remember the fact that spring, the longitudinal bending, is always towards the bark, the problem of coping with this characteristic can be much alleviated. A common domestic commodity which suffers from the same propensity to spring when cut longitudinally is celery.

In many hardwood logs spring is not a problem, but in an important percentage of logs it can be sufficiently troublesome to cause downgrading, and, in some cases, rejection, of sawn products. When spring is a problem, the products should be backsawn rather than quartersawn.

Remembering that the longitudinal bending always takes place towards the bark, we can see from Figure 6.5

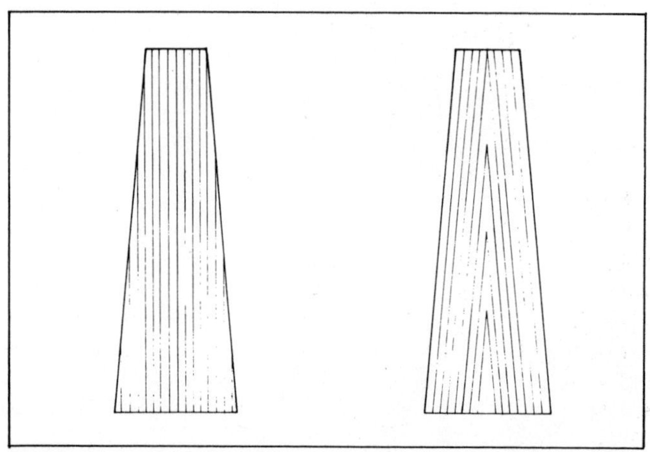

Figure 6.4. Diagram illustrates how taper sawing produces shorts in centre of log.

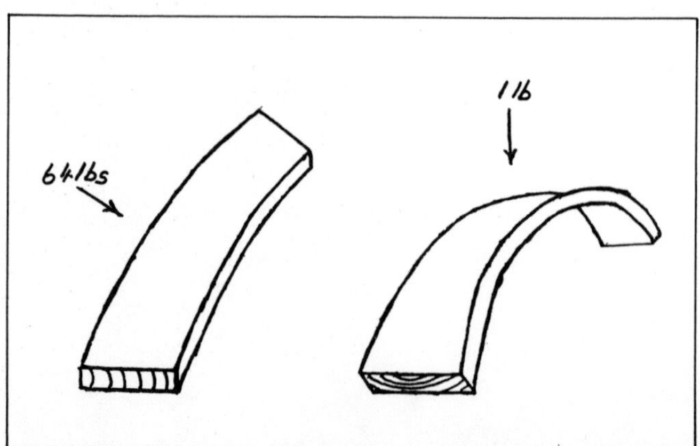

Figure 6.5. Illustration of direction of distortion caused by longitudinal growth stresses in both backsawn and quartersawn products.

that if the product is backsawn the spring of the log results as bow in the sawn product, while if the product is quartersawn the resultant distortion is much more difficult to contend with. An indication of the degree to which the product is amplified by quartersawing is given by the fact that if the sawn product in the figure is 4-inch by 1-inch in section and if a 1-pound force is required to straighten the backsawn piece, then the force required to straighten the quartersawn piece exhibiting the same amount of distortion, would be 64 pounds. In the figure the degree of distortion has been purposely exaggerated to make the point more obvious.

Unfortunately some important buyers of dressing quality boards prefer up to 85 percent of their orders to be quartersawn because of the more attractive ribbon figure. This applies particularly to the merantis with interlocked grain.

When flitches spring during resawing there is a problem of subsequently producing pieces of uniform thickness from the bent flitch. Therefore, the sawing equipment must either:

1. Allow the flitch to be "face cut" to straighten it, that is, straightened by a cut made "by eye" and without the flitch against a gauge or linebar, Figure 6.6, or
2. The equipment should force the flitch straight and hold it that way while it is sawn.

The first requirement necessitates a sawing machine capable of passing the flitch across the bench with only the driving force provided by the horizontal feed rollers and/or the sawyer, and without any assistance from vertical feed hobs pressing the flitch against a gauge.

The second requirement is provided by a combination of carriage and linebar gauge. Flitches are loaded onto the carriage over the linebar and are pushed against it by knees, each of which is independently operated but which, after assuming the profile of the off-saw side of the flitch, can be locked into that position relative to each other, by clutching to a set shaft.

Also we have found that dogging logs or flitches within two feet of their ends on a sufficiently robust carriage provides a reasonable chance of holding the log or flitch against its tendency to spring. Consequently, carriages have

been designed on which one end knee is moveable longitudinally along the carriage, to permit a range of log lengths to be dogged as advantageously as possible.

Further, a range of equipment is available which, for small logs, allows pieces to be removed concurrently from either side of the log, thus preserving the balance of growth stresses in the central flitch, thereby reducing its tendency to spring. Such methods include gang framesaws, two saw log edgers employing circular saws and the well-known twin and quad bandsaw systems, incorporating a range of feeding devices and sometimes employing chipper canters. While such systems alleviate the problems caused by high growth stress gradients, they suffer the distinct disadvantage, as far as tropical hardwoods are concerned, that they do not permit the log to be sawn for grade according to the log quality revealed as sawing proceeds.

Once again, there is no way of quickly and reliably determining before sawing commences whether or not a log will spring to a troublesome extent, so that the layout of the sawmill must permit flexibility in choice of sawing system. For example, flitches containing high growth stress gradients cannot be profitably sawn on multi-saw edgers, because distortion can occur in the final products.

Multi-rip sawing to width of wide quartersawn flitches from trees containing high growth stress gradients, results in the sawn pieces bending as they are produced, according to the stress distribution in the flitch. Such spring can be enough to render such pieces unmarketable at higher grades.

Shakes

Another result of high growth stress gradients is the star pattern of splits or shakes that sometimes occurs on the ends of the logs. Obviously grade recovery is improved if the sawmilling equipment permits such logs to be turned rapidly so that the prime saw cuts can be made predominantly along these shakes. In this way, flitches for resawing are produced with the shakes on the edge of the flitch rather than across its cross section.

Profitability must be improved when the shake is located in the flitch so as to be confined to one final piece when the flitch is ripsawn, rather than be located across the flitch, thereby affecting a number of the final products.

Seasoning

Most species of tropical hardwoods are relatively easy to season, except that when kiln seasoning for export markets in which the equilibrium moisture content is low, the kiln design and drying schedule should be carefully chosen to avoid large variations of final moisture content throughout the charge and also to avoid degrade, particularly surface checking. Surface checking can be a serious problem when drying the thicker sizes of species such as ramin, kapur, merbau and keruing, unless treatments such as microcrystalline waxes are employed or, better still, these thicker sizes are quartersawn, thereby restricting checking to the edges of sawn pieces. Also, some timbers with interlocked grain, such as some of the merantis, are preferred to be quartersawn by some buyers because of the more attractive ribbon figure. This applies particularly to the Japanese market.

Therefore, once again the sawmill system adopted should be sufficiently flexible to permit both backsawn and

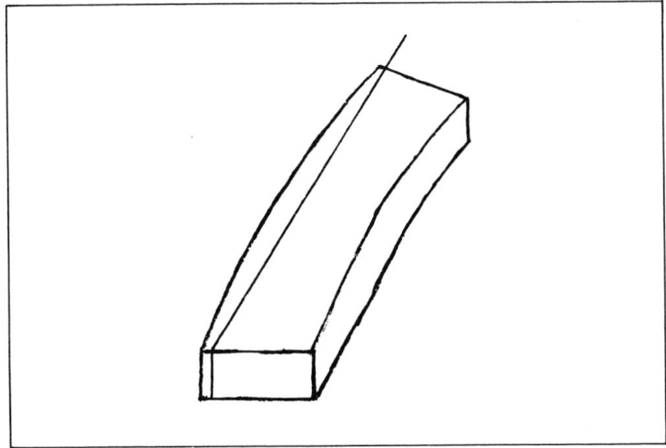

Figure 6.6. Illustration of how sprung flitch is straightened by face cutting.

quartersawn to be produced at will, and you all appreciate that quartersawing requires considerably more turning of flitches during conversion than does the simple through and through sawing systems for backsawn production.

Recommendations

Throughout this chapter I have stressed the need for production flexibility. Another description of what I have been talking about is to call it "grade sawing." For any mill sawing a range of tropical hardwoods, the problem is one of grade sawing rather than simply sawing through and through, and I believe that for the near future conditions in Southeast Asia this requirement is best met by a combination of:

1. A powered log deck with fast and accurate turning gear capable of handling a range of log sizes. Preferred system incorporates log kickers, Simondson type hook turner for large logs and chain type nigger for turning small logs before sawing.
2. Log carriage with facilities for taper sawing, and incorporating flippers for turning down against log deck kickers.
3. Band headrigs of as large a wheel diameter as can be afforded and preferably with short columns, so as to avoid vibration problems (Figure 6.7).
4. Band resaws capable of face cutting.
5. Multisaw edgers for those species not prone to spring.
6. Small recovery benches.
7. A defect docking system handling only those pieces actually requiring docking.

In mills of certain sizes, facilities for storing flitches and reloading these back onto the carriage for further sawing can also be an advantage. Also, if residues are to be converted into pulp chips, the provision of a headrig chipper in front of the band headrig should also be considered. One example of a suitable arrangement of the equipment suggested is as shown in Figure 6.8.

This mill, which I stress is only one possible arrangement of such equipment, and in this case is for a relatively small input of a range of log qualities, consists of:

1. A log transfer.
2. A log deck incorporating fast loading and turning gear capable of handling a complete range of log diameters.

3. A remotely controlled log carriage with a relatively simple sizing system, insensitive, as far as possible, to changes in input voltage, a rapid taper adjustment on head blocks, pneumatic dogging, pneumatic flippers for absorbing log loading shocks and for rapid turning during sawing and a manually operated receder.
4. A normal strain band headrig, with wheels as large in diameter and as wide as economically possible and built with short sturdy columns.
5. A discharge rollcase, providing access to other machines and also a direct outlet out of the mill.
6. A flitch storage and reload deck.
7. A powered transfer from the headrig discharge rollcase to a band resaw.
8. A band resaw and roundabout system, operated by two people, incorporating a linebar and twin radial arm feed hobs, and capable of "face cutting."
9. Transfers from the headrig discharge rollcase to a multisaw edger.
10. Multisaw edger, with the number of moveable saws according to maximum log size and with one fixed saw and a moveable linebar.
11. Transfer table across back of mill, providing facilities for receiving material from all machines, discharging waste from both band resaw and multisaw edger, feeding pieces from both band resaw and multisaw edger to multisaw edger for further processing, directing pieces to docker as required and feeding to a sorting system.
12. A central, undermill waste conveyor, receiving waste direct from band resaw roundabout system, the transfer table, and via a conveyor belt, from the defect docker.
13. A defect docker system, fed from the transfer table and receiving only those pieces actually requiring attention. Seasoning material should, if possible, be end-trimmed after seasoning.
14. Waste discharge skids for decayed or brittle heart material and other waste originating from the headrig, fed from headrig discharge rollcase and cleared at regular intervals by forklift.
15. Transfer table to a sorting system, which could be either green chain, circular table or edge sorter.

The arrangement I have shown is only one of many that are possible and I have used this layout simply to illustrate the principles that I believe should be incorporated in mills designed to handle the general run of the Southeast Asian hardwood forests.

Figure 6.7.   Heavy-duty band headrig and log carriage suitable for tropical hardwoods.

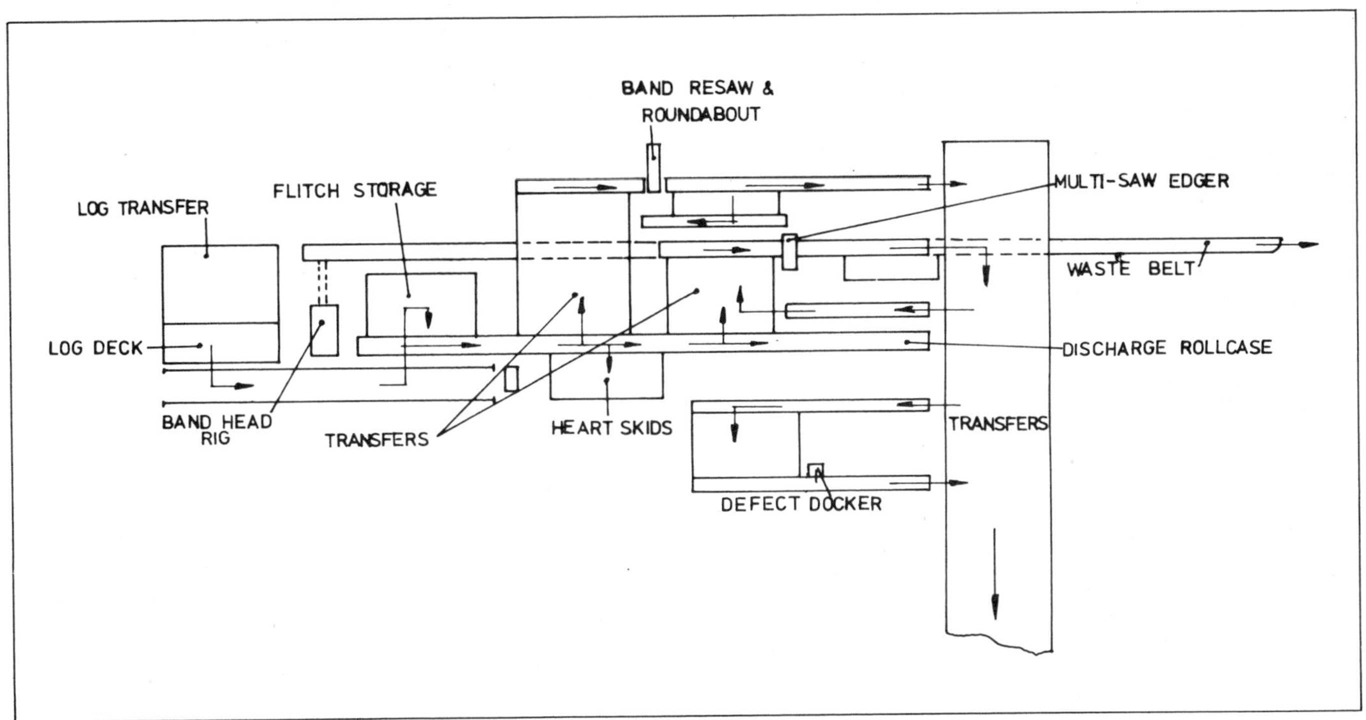

Figure 6.8.   One suggested layout suitable for milling tropical hardwoods.

CHAPTER 7

Training, Utilisation, Development, and Balance

MAURICE C. TAYLOR
Managing Director, Southern Cross Engineering Co. Ltd.
Christchurch, New Zealand

My job is the designing and development of sawmills in both soft and hardwoods with a particular emphasis on the smaller units, though I should say at this stage I am not against larger units. However I believe that in the smaller field the problems are more personal and the need greater.

As I have to eat, I also run a company, and we make about 250 items of sawmilling machinery. I am known as Mauri throughout New Zealand and in many places overseas and get on well enough with most people. I have a particular interest in Southeast Asia where the need for teaching at the lower levels is very great.

I have called this chapter "The Four Principles" with the principle of training in the priority position. Development and improved utilisation will be a natural follow-up from this and are, in fact, part of the training programme. The fourth principle, "balance," is a separate issue but has to be considered with the others.

Where there are joint efforts already in operation or projected (and by joint effort I mean participation by an overseas company in finance and management as well as operation), no help from me is needed. Nor is the problem, generally speaking, at the top technical or management level, except that top people sometimes show a singular lack of perception in the selection of machinery and plant layout design, not always taking into account local conditions and abilities.

Training management

In earlier and less complex times, promotion usually came from the bottom up. A man got a little promotion, acclimatised to it and then went a bit further, and if he was good enough he finally got to the top the hard way. Many of the world's major businesses started out in this fashion and many a managing director sat in the driving seat, may I add successfully and competently, with qualifications that would not get him a junior clerk or floor manager's job in these times.

Conversely, quite a significant number of graduates today expect after leaving school to start near the top and most manage to start above the area of practical routine work. This leaves a gap in the vital area where the money is earned.

In New Zealand and, I believe, several of the Western nations, we have this problem of the gap. Skilled operators and supervisors are at a premium in spite of the fact that we have a good education system and good apprenticeship schemes backed by technical institutes. The situation is much more difficult in a developing country as yet lacking the necessary organisation to develop these facilities.

Normally, of course, time will solve this problem but in Southeast Asia, time is something we do not have. Outside pressures make a rapid urbanisation and the developing of manufacturing industries a vital necessity so that work may be found for the masses and the standard of living may be raised. Failure to do this will almost surely have far-reaching results for in today's changing world the future of all will depend upon the comfort and goodwill of the many.

Practical assistance is of far greater use than gifts of money. The giving of money to a man, however well intended, reduces him to the level of a beggar—he will hate you for it. But help him to earn it for himself, and you earn his gratitude.

In referring to timber development, I will deal with emphasis on the sawmill mainly because sawmilling is my business, and it is therefore an area in which I have some qualification. It is also the first stage in the processing chain.

Obviously if we are to short circuit time, outside aid is required. In can come in three forms—

1. From private overseas companies.
2. From overseas governments.
3. Jointly by an overseas government aided by private industry in that country.

To illustrate an example of an aided project, I had the good fortune to be associated with the development of a sawmill unit in West Irian Indonesia which was financed by the United Nations Food and Agricultural Organization. It was a simple setup and it has been so successful that we are now to enlarge it. It is a great pity that the area in which it is installed is so isolated that it was ruled out as a general training centre, but it has provided valuable local industry and local training and is now locally managed.

As far as private overseas companies are concerned, the giving of aid is more difficult. They have shareholders who want dividends, so they must make a profit. Participation in the form of joint ventures partly fills the bill, but only partly because it is of necessity limited to the immediate operation, and we are concerned with general training. They must also safeguard their interests and to do this they must have a large measure of control in such a venture.

Training carried out in these ventures will eventually filter outwards but this is not quick enough for rapid development, and so something further is needed to provide general training open to all. Therefore we turn to government sponsored training schemes.

Should an overseas government sponsor a training scheme outside its own border, there must always be the possibility that this assistance is not without political bias. It should of course be the aim of local government to establish its own training unit with outside advice and some practical assistance in the initial stages.

Once established it should, I believe, be under local government control and I feel sure that local industry will and should support it. So far it is somewhat vague so I will try within the limits of this chapter to be more specific.

New Zealand government training scheme

Our Government Forestry Department maintains and runs a sawmill training school in New Zealand. This school teaches saw sharpening and maintenance, a vital part of sawmilling. The small sawmill is equipped to teach band-sawing and circular sawing methods and it can demonstrate most sawmilling operations.

However it is not equipped to run under commercial conditions and cannot, therefore, round off its instruction. This need is not so important in New Zealand because we have many quite well-run operations capable of giving, as it were, the commercial working instruction. But in Southeast Asia I feel this instruction centre must be able to take instruction into commercial cutting for *profit as profit is the name of the game.*

I have heard it said that we should at this time go back to the beginning. I do not believe this for backward steps are not progress. What we must do is to use simplified modern machinery to teach basics so made that we can give advanced instruction as knowledge is gained. The training sawmill must not be too big, or the machinery too complex, for it must, as well as instruct in new techniques, be able to show the way to improve standards on the large amount of equipment operating in the area already.

Machinery requirements

Regarding machinery, a good bandmill is essential and my experience of the industry tells me it has got to be a good quality heavy weight machine using wide blades of average gauge and average tension (Figure 7.1). Power lift to the tension bars is essential, as is powered pressure guide, but no gimmicks—just a basic machine that will stand up to hard work, cut good timber even when saw doctoring is indifferent. I would want to use the bandmill to break down logs, to show how it can be used to recut part sawn timber, how it can even be used to cut accurately finished boards. In other words, how to get the most out of it in a big mill or a small one.

The carriage again needs to be a heavy one. It is necessary to have good robust tong dogs with two positions and automatic pull back which ensures that the log is always pulled firmly against the headblocks, for how can we cut accurately if we cannot hold the log firmly in position? (Figure 7.2).

The carriage must have power drive to the log feed coupled to an accurate indexing mechanism, but the main requirement will be simplicity and ease of understanding for we are teaching basics. In passing I would mention that my test for accuracy on a bandmill and log carriage is to set the indexer to 2 millimeters and take off the log unbroken, a board 5 meters long by 2 millimeters thick, and that's accurate sawing.

Manhandling timber is hard work. It's wasteful of manpower and slow, so we must have a good powered turning system remotely controlled by the sawyer capable of demonstrating all types of turning and, above all, to turn efficiently without damage to the timber or to the equipment (Figure 7.3). Good mechanical handling is a must because if it's hard work or difficult the operator just lets it go and poor quality cutting and increased waste is the result. This is what we are trying to avoid.

Behind this main saw we would propose a band resaw (Figure 7.4). Again, simple and uncomplicated, but we will take the hard work out of it and that means power feeds. We need accurate cutting and that means an indexing set-works. The machine needs to be versatile in its application and that factor will decide the type of resaw to be used.

Firstly, the machine will have, as far as the saw is concerned, the same characteristics as the bandmill breaking down unit, and we would fit it up to cope with a large variety in cutting patterns. We want to be able to cope with round backs, that is, a timber with only one flat face so we must be able to use the machine open. We also want to use it to cut timber with a rough edge on one side to produce square timber by sawing wide boards from one edge. This procedure illustrates the bandsaw's greatest advantage and that is deep sawing. We also want it to edge wide boards flat down. We also want to demonstrate this saw's proper function and that is the fast production of boards from part-sawn timber or accurately presized squares.

It's a tall order, and to achieve it we will fit this machine out as a stub line bar. It will have a fence 4 feet long. Independently driven hobs maintain a firm pressure against the fence. Feed speeds will be fully variable. It will demonstrate just what can be done in a well designed heavy weight machine of this type. Open it wide and it will do simple cutting.

With care these two machines can be used without any other machinery and used to their full capacity we would expect an output of sawn timber per 8-hour shift of about 40 centimeters per shift which is much more than is being done with much more machinery in many places in Southeast Asia today.

A third machine needs now to be introduced into this sawmill. This needs to complement the other two to share the load and this can be done one of two ways. We either put in a small band resaw or circular saw to salvage the indifferent timber, and we leave the larger resaw to cut larger quantities of finely finished timber at speed, or we can use the bandmill to break down logs, reload them, and cut finished timber, using the resaw to salvage and to edge wide boards (Figure 7.5).

In most cases the edger will be a better machine than the small band or circular resaw, particularly if we use the type that will edge both flat boards and round backs. The rest of the mill consists of various conveyors designed to move the timber around at all stages mechanically.

Edge trimmers and a sorting table will be provided, that is, the bits that go to make this small compact sawmill. Let us now look at it as a whole.

It can be operated at different levels of production using different machines or as a whole integrated unit operating on a continuous breaking down system, where each machine forms part of the chain rather than one breaking down saw supplying several similar producers of finished timber.

It can demonstrate and instruct on a great number of techniques and varying sawing patterns. It is simple enough to give valuable training in simple sawing but it is also sufficiently equipped to teach modern methods and, most important of all, to operate commercially to show what it means to produce a good quality product with minimum waste in commercial quantities with a profit.

I have at this time over 35 sawmills of the basic type and design shown in Figure 7.6, with of course variations in both hardwood and softwood at work. All are operated by small companies, the labour largely unskilled or semi-skilled, most of whom are trained on the spot. The greater part of the key personnel have at one time or another taken courses in our government school at Rotorua, New Zealand.

Finally, this pilot training plan is designed to teach technical skills, machine operation, saw doctoring, machine maintenance, sawing techniques, grade recovery and conversion, and management at mill floor level. It is going to need some very dedicated expatriate people for a period until local people can take over. In fact, the first job will be to teach the teachers but where there is a will to do this it can be done. *It has to be done.* In fact it has already been done elsewhere. It is being done successfully in New Zealand and the number of people that come to New Zealand for our practical course from the Pacific Basin is surely recommendation enough.

The unit at Djajapura, West Irian, with which the writer was associated, is a further good example of what can be done in the most primitive of conditions where there is the will to do it and the dedicated people to start if off and stay with it as long as is necessary.

Our New Zealand government has now voted the necessary monies to support the establishment of a sawmill training scheme at Kuantan, Malaysia, which will come under the control of the Malaysian government body, MARA. It will be built in large part from New Zealand-made machinery. Initially it will be staffed and run by New Zealanders who will teach the necessary engineering and practical sawing skills. An important part of this project, however, will be the training of local sawmill instructors who will take over when the New Zealanders leave.

The training sawmill layout I have described in this chapter forms the writer's submission for the MARA Project. It will have the right sort of dedicated people that were found for Djajapura who I know will have the necessary will to make it a success.

The foregoing statements assume that training is the responsibility of government through the tax systems or by levy from the industry. Some recovery of funds should be possible through fees but these must be kept low or the service may not get used in the very areas where it is needed most.

Where direct aid cannot be given to establish a training sawmill it is possible to arrange long-term low interest rate loans. On this basis a private company could establish this training mill, operate it commercially and offer to train as well on government subsidy. It's a practical way of bringing governments and industry together.

The importance of balance

I come now to the latter part of this chapter which is concerned with balance. Experience in New Zealand and elsewhere tells me that even in a developed industry there is room for small and medium size as well as large industries but the term small is a relevant one and can be interpreted differently by different people.

It's a matter of comparison and the standards which are set by local conditions. If all you have ever owned is a bicycle, the next logical step is a motorised bicycle. If you go straight from a bicycle to a 150-mile-an-hour motor car, it's pretty certain that you will break your neck and possibly several other people's as well. If all you have ever successfully owned and run is a small primitive sawmill and you go straight to a highly modern efficient unit you may not break your neck but you will lose your money as will anybody else who has been foolish enough to support you.

I have heard it said in Southeast Asia that the type of ideas I am trying to convince you are correct will not work without overseas participation and not always then. The fault is not in the machine because it is an inanimate thing and only comes to life under the control of its operator who has to learn how to handle it and control the business of which it forms a part. This train of thought is not solely confined to Southeast Asia. In my own country, my personal efforts to try and modernise small sawmilling met much the same reception. Following is a true story.

The South Island of New Zealand is divided by a chain of very high mountains. On the eastern side we have low rainfall and excellent conditions for growing softwoods. Our growth rate is one of the fastest in the world. On the other side of the mountains we have a very high rainfall (up to 130 inches per year) with tropical vegetation and hardwood trees which can run up to 10 feet in diameter. Some species take several hundred years to mature. The timber is very dense and cutting conditions would be, I think, worse than you would have in Southeast Asia.

In my early efforts to convince small sawmillers that they must change to new methods, I was told by the softwood millers that it couldn't possibly work as the saws would run around the knots in the timber instead of going through and that I should go to the West Coast where the hardwood timbers were straight without any knots and should be all right for this method of cutting.

The hardwood sawmillers were equally positive—use of these saws and these fast methods of cutting were not possible, and I should try and convince the sawyers of the softwoods on the other side of the mountains because the timber over there was only rubbish and would pose no problems.

One of the statements made to me by hardwood and softwood sawmillers in New Zealand in their efforts to convince me these ideas were wrong was that although they admitted that the large sawmills were successful using these ideas, they could not be scaled down, and in any case large mill operators were wasteful because nobody had any time to cut grade.

I stopped my car at the top of the mountain pass between the two areas and I pondered a little, and I was sure that I was right. I have no time within the scope of this chapter to tell you of my early problems when every success was minimised, and every little failure magnified.

However, 33 sawmills later with the old methods largely superseded in both hardwood and softwood cutting, I could safely say that the case is proven. It took about eight years to achieve this in New Zealand, I repeat, eight years.

I have heard plenty of people in Southeast Asia say they are going to revolutionise the industry overnight. They have got to be joking, or they don't realise what they are up against. Their biggest job is to convince man that he has to change his ways, because the mass of people are conservative, irrespective of their political views. They don't like change and the older they get the less they like it. They are afraid of it and therefore they obstruct it.

It is essential even in the early stages of development that local people should manage their own affairs and this can be done by either starting separate smaller sized ventures locally run as well as joint ventures with overseas participation or by taking existing units and upgrading them to the desired level.

If we are to consider the upgrading of an existing business, no matter how poor the original setup was, there is already something to work with—site, some people, some equipment, and a business in existence. I know that many of these smaller units could not be economically upgraded, but I also know that there are some that can.

As this is in the private sector, straight-out aid from an overseas government is unlikely but plenty of advice and practical assistance are available, and as far as New Zealand is concerned, there are possible long-term low interest rate loans available if they meet the authority's approval.

There are private individuals within the industry who are prepared to give advice in this direction and some firms in the supply field, including mine, which can offer a complete turnkey project service, including teaching, installation, and commissioning.

Loans can be obtained on the installed cost of machinery so that although these services have to be paid for, a loan may be granted for them as part of the overall project. The training sawmill I have described would also make an ideal basic sawmill for private enterprise. It is capable of producing a reasonable amount of ouput and a good quality product. It can be so laid out that future development is possible as skill and techniques in operation are improved.

It can be put together to suit an exact local need and because successful small businesses invariably become successful medium-sized businesses and then successful large businesses, this type of development has considerable merit. People grow with it and this is the balance which is mentioned at the beginning of this chapter.

Balance means that in any particular field there are small, medium, and large undertakings *not* working in opposition but complementary to each other. It can be said that, in the main, large ventures will be operated jointly by local and overseas interests for a period in Southeast Asia for obvious reasons. There could be some overseas participation in medium sized ventures and there can be small businesses which are entirely locally run and it is these that will provide for tomorrow.

The big businesses will probably cater mainly to an overseas market; the medium sized businesses can cater to some overseas and some local market; and the small businesses will cater entirely to a local market. *But* a number of small businesses pooling their resources can successfully serve major overseas markets, and as the use of the local raw materials become more common in their country of origin, the very large undertakings will suddenly discover that they have a very attractive market right on their own doorstep.

I have lived and worked in large industrial countries. My later years have been spent in New Zealand which is a small country of three million people. In New Zealand, I learnt one of the most important lessons in my life and that is the value of the small and independent country or business which can think and act for itself.

Figure 7.1. A good bandmill is essential to an efficient sawmilling operation.

Figure 7.2. The log carriage should be heavily constructed for best operation.

Figure 7.3. A good log turner should operate without damage to either the log or the equipment.

Figure 7.4. The band resaw should be uncomplicated, but with power feeds and indexing setworks.

SAWMILL TECHNIQUES FOR SOUTHEAST ASIA

Figure 7.5. An efficient edger should edge both flat boards and round backs.

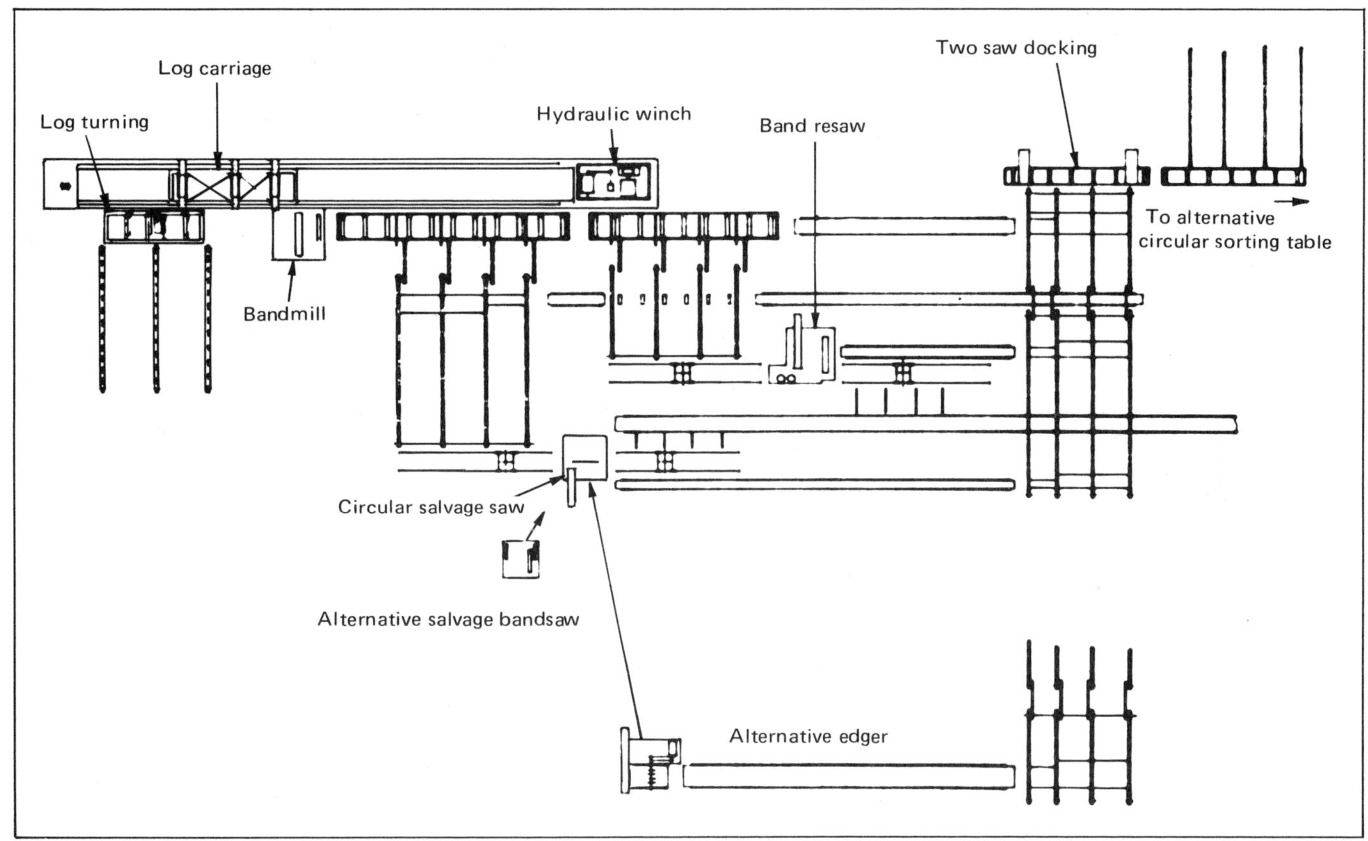

Figure 7.6. Typical sawmill layout for operation by unskilled or semiskilled labour.

CHAPTER 8

Portable Sawmills for Heavy, Hard-to-Reach Logs (Part I)

BRUCE A. WICK
Director, Wick Industries International
Vancouver, Washington, USA

I think that I shall never see, a poem lovely as a tree.
A tree whose hungry mouth is pressed,
Against the earth's sweet flowing breast,
A tree that looks at God all day, and lifts her leafy arms to pray.

I believe most of us are familiar with this famous poem and I'm sure we all agree that trees are one of man's most precious resources. I have personally seen trees left to waste and rot in North and South America as well as in the Far East where I've spent the majority of my time these last eight years.

I do not think anyone wants to see this vital resource wasted. This waste is costly to our countries as well as to our own personal pocketbooks. When you consider the time, energy and money it takes to build roads and infrastructure and the costs of cutting down a tree, only to find it does not have enough recoverable wood to afford hauling it out of the forest profitably, you must stop and ask yourself, "Is there a way this resource can be recovered?"

The answer is "Yes!" The way is to take transportable sawmills into the forest where the trees are being harvested. By doing this, you can save up to 60% of your log trucking costs, reduce the cost of your sawmilling operation, and earn back the money you've spent getting to and harvesting these trees. Cutting logs into cants or dimension lumber at the landing eliminates all wood waste at the landing. Even the good wood in cull logs, which normally would be left to rot, can be recovered. Thus, you are able to recover the maximum amount of wood in any stand. And because only usable wood is hauled out, trucking costs are then reduced by as much as 60%. In addition, a central, permanently located mill now need be only a resaw operation instead of the normal complete sawmill. This can result in a tremendous savings in overall costs.

Transportable sawmills are specifically designed to be taken into the forest, in the toughest logging shows, wherever they might be in the world, to efficiently cut any species of log, from the softest to the hardest.

Encountering problems

The mill in Figure 8.1 is located on the Batu Islands, off West Sumatra, Indonesia. The problems encountered here were manyfold. There was absolutely no infrastructure, or even a town or village where workers could be drawn from for employment. This meant that before any operations could be started a base camp had to be built. This required lumber and there was none within the area.

Capital equipment consisted of one D-6 tractor and one transportable sawmill. The only way to get this equipment to the area was to transport it over water on a pontoon to the edge of the forest. The next step was to cut a path through the mangrove swamp, bordering the forest, to solid land above the high water mark (Figure 8.2). Not having any handling equipment, it was necessary to disassemble the mill and carry it by hand to the sawmill area where it was reassembled. This task took three days and required twenty men. During this time, the tractor had skidded a number of logs which were ready by the time the mill was ready to start sawing.

The logs had to be debarked by hand and it took three men to keep up with the sawmill (Figure 8.3). Skid logs were placed under the mill, extending some 200 feet beyond, so that the tractor could fill the skid ramp for a full day's sawmilling production (Figure 8.4). We used a hand winch to pull the logs along the skids and under the mill for sawing. It was not very fast, but it was effective during the first stages of operation (Figure 8.5).

We had considerable rain that first week, and even though weather conditions were miserable, we managed to saw enough lumber to build a longhouse to house the men and a tool house. The tool house was erected on top of the pontoon which was used to transport the mill and tractor to the area. We filled the pontoon with water and pumped this water to a series of barrels on top of a platform erected in front of the tool house. We were able to continuously supply water to the sawmill blade through a plastic hose.

Our next step was to build a pier facility so that ships could be anchored alongside for loading. A heavy block of hardwood was mounted on a wooden frame with a rope pulley and used to drive piles for the pier. We sawed the decking with the mill (Figure 8.6).

Transportable sawmills are generally simple to operate and maintain. Our sawmilling crew for this project was able to operate the mill after three days of training. It took 90 days to complete the pier, which was able to store more than 4,000 cubic meters of sawn timber, which in this case, was railroad ties destined for the Java, Indonesia, market.

The first shipment of some 3,000 cubic meters of railroad ties, or sleepers as they are sometimes called in Southeast Asia, was made approximately 160 days from the time we landed on the beach. We installed a light gauge railway directly behind the mill and down the full length of pier. The ties came directly off the mill onto a hand dolly which transported them along the pier for storage and later to shipside for loading (Figure 8.7).

I must admit, this operation was one of the most difficult and challenging, but it was one of our most satisfying.

At a mill installed in Aceh, North Sumatra, Indonesia, we first set up at the camp site and logs were hauled to it. At this site, we manufactured dimension lumber to build housing for more than three hundred men. While this was going on, a road was being constructed to the forest area. The road construction was progressing smoothly until the crew encountered swampy terrain. They tried to use round poles as a road base over the swamp, covering these with rocks and gravel. This system did not work because the poles rolled beneath the gravel when the trucks moved over the road. It did not take long before the road developed holes and became unusable. The manager of this company had logged for many years in the United States and had constructed a lumber-based road previously. He decided to try this method in Aceh and the transportable mill was moved onto site. Trees from the road right-of-way were sawn into 3x10 and 3x12 inch planks, 26 feet in length. These were nailed together and used as the road base. Gravel was spread over the top. More than nine kilometers of this type of road was constructed, and it has held up to continuous use by log trucks and oil drilling rigs for more than three years. The transportable mill was moved continuously during this construction and the crew was soon able to take it down and reset it up in less than three hours.

After the road construction was completed, the mill was moved to the company's loading area. Logs were rafted and floated down the river to the sea for ship loading. More than 30% of their logs were sinkers which made rafting a nightmare. A crane was used at the loading site to unload trucks, place the logs in the river for rafting and to feed the mill.

If a log sank, it was placed under the mill for sawing. The local market kept the mill busy two shifts a day. A second transportable mill, this one an electric-powered model, was installed with a complete remanufacturing plant. From then on, the two transportable mills were utilised to break down the logs into cants for the resaw.

From time to time, specialty items were called for, such as solid table tops, four feet wide by twenty feet in length. Large and long bridge timbers are another product that was produced with transportable mills.

A mill has been operating in Sabah, East Malaysia, for more than four years. The company that purchased this mill had a unique problem. They had logged over one million of selangan batu that they planned to saw in their conventional band mill. You that are familiar with hardwoods in Southeast Asia, know this species to be very durable but also very difficult to saw. It didn't take this company long to realise that they had embarked on a very difficult and unprofitable venture. Their mill was just not capable of sawing this species.

It looked as if they had no alternative but to abandon these logs as the only market was for sawn timber. It was then that the company decided to try a transportable sawmill. It was the right decision as production was high, costs were low, and they were able to recover 100% of the timber felled.

Selangan batu has been used for railway ties in East Malaysia as well as the Philippines for over a quarter of a century. In the past, the ties were mainly hand-hewn. There are numerous other species that have never found a market because of the problems encountered in sawing. Transportable mills have changed this as they are capable of economically sawing practically all known kinds of timber. Selangan batu, known as yacal in the Philippines, is one of the most expensive construction lumbers in their country.

When to use transportable sawmills

When does it make sense to transport the sawmill to the logs rather than bringing the logs to the sawmill? The company that owns this mill didn't have to think twice about this answer. Their timber concession is more than seventy five miles from their loading point. You don't have to transport logs seventy-five or fifty or even twenty miles before it makes sense to take the mill to the logs.

Today, the cost of transportation is one of the highest costs found in the timber industry. When you are hauling sawn timber as opposed to round logs, you need half as many trucks, use less than half as much fuel and half the maintenance and manpower.

Far up the Kayan river from Tarakan, East Borneo, is a mill sawing for local as well as the export market. The company operating this concession is one of the oldest and the largest timber companies in the Philippines. They own and operate a pulp and paper plant along with a sawmill and plywood factory. They are very much aware of the value of all species and utilise practically all of the tree. I would venture to say, their motto is "Waste not—want not."

Generally speaking, portable mills are relatively easy to transport. If roads are available, they can be hauled on a truck and trailer. Where this is not possible, they can be dragged on skids behind tractors. Set up time is usually short. Most types can be set up in a matter of a few hours or at the most, one day, without the aid of auxiliary equipment. After four years of daily operation, this mill was joined by a second unit. They are moved from log landing to log landing after the export logs are removed. Both are operated by private contractors hired by the company. The first mill was one of the only sawmills in Balikpapin five years ago.

Like so many transportable mills, these mills were first used to saw lumber for the construction of base camps. These particular mills have been instrumental for building housing, schools, a hospital, recreation facilities and infrastructure for more than ten thousand people.

Last year, we installed a mill on Samar Island of the Philippines. Due to the close access of timber, it was set up in conjunction with a conventional sawmill. Hard-to-saw and reject logs were causing a low output in the main plant so the transportable sawmill was brought in to break down these types of logs for the resaw. Almost immediately, production in the conventional mill was increased by more than 30% due to the fact that the head saw cut only solid logs. A live green chain has been installed to transfer the cants directly to the resaw. The company recently told us they planned to bring in a second unit for their woods operation.

On Sumbawa Island, East of Bali, Indonesia, a mill saws reject logs for parquet flooring. The company was able to accumulate a shipment of over 3,500 cubic meters within 75 days after the mill was installed.

A number of countries throughout Southeast Asia have passed laws requiring concessionaires to manufacture their timber into semi and finished products such as baby squares or cants, lumber, railroad ties, plywood, and veneer. These laws were passed to encourage more local manufacturing, create additional jobs within the country and earn the additional foreign exchange. Canada has had such a law for many years. They call it a primary manufacturing law.

To develop conventional mills and train operators requires too large a capital outlay for many smaller companies and would take far too long to fit into the government's timetables for phasing out the export of sawmill logs.

An excellent alternative to conventional sawmills is transportable sawmills as they cost much less, take less time to train personnel, are easier to operate and maintain and increase overall production in any stand of timber. However, there are problems associated with transportable sawmill operations. For example:

- Marketing — the producer must be able to furnish quantities, sizes and grades acceptable to the buyer.
- The product must be competitively priced.
- Shipments must be large enough to encourage ships to call.
- Shipments must be made within a reasonable time after receiving the order.
- The lumber must adhere to a set of recognised grading rules.

Few firms in Southeast Asia have the resources and size to enable them to meet all these vital requirements. For every large firm, there are hundreds of smaller firms without the required capacity for this market.

How, then, can smaller companies expect to share in this fast-expanding export lumber market, utilising transportable sawmills, when they don't have the size to compete in the export trade?

Most forest areas have large quantities of exportable timber as well as minor species not acceptable on the export market. Therefore, the export program should concentrate on those species which are known in the market place, while the minor species should be sawn for the domestic market.

A concentration yard should be set up to enable operators of transportable sawmills to successfully compete on both markets with the giant firms. The concentration yard should be export-oriented initially, and later turn to the domestic market as the buying public accepts the minor species.

The transportable sawmill located in the concession area will have to saw a variety of species and therefore will not be able to produce enough of any one species to fill an export order. By combining the production of a number of transportable sawmills, a quantity of one species can be built up faster at the concentration yard for an order.

Installing a resaw at the concentration yard enables each transportable sawmill to saw larger pieces, which greatly increases production, decreases waste, and enables the concentration yard to saw for specific orders.

A good concentration yard can accommodate the production from up to ten transportable sawmills. Grading, bundling, and shipping can then be done at one point rather than duplicating these operations at a number of sites. Therefore, less expert personnel is needed and costs are dramatically reduced. Also, quality control becomes an easier task.

By having a number of transportable sawmills producing for a concentration yard, you also have a built-in safety factor. If a major piece of machinery breaks down in a conventional mill, usually production must stop. Often, this creates a shipping delay, and export buyers are generally unsympathetic to this problem as they usually have the order committed for a certain delivery date at a specified price. Also, shipping usually has to be rearranged and this becomes costly to everyone involved. If, for example, you have five or ten transportable mills in operation and one breaks down, you only lose a percentage of your production. This loss can be made up by operating the remaining mills overtime.

It is much easier for a concentration yard to maintain contact with the export market than it is for a number of mills operating inside the concession area. It is also easier for the buyer to maintain contact with the seller if he can reach him at a central location. This system requires less manpower and expense for both the buyer and the seller. The savings are to the seller's advantage.

If logs are easily accessible to the concentration yard, it sometimes pays to set up the transportable mills at the yard instead of in the concession area. The main advantage for doing this is that the mills can usually be worked longer hours and thus production is increased. The main drawback is that waste will be accumulated and disposal can become costly as well as time consuming.

You may ask if the export buyer is receptive to buying semi and finished timber products as opposed to round logs. If you would have asked this question a few years ago, I would have had to answer "No!" Today, I can definitely say "Yes!" With the high cost of shipping, pollution control, and labor in most buying countries of the world, the buyer cannot absorb the costs of handling, shipping, and disposing of waste and still sell at a price that the market will buy. I, for one, think it would be tragic for us who are in the timber business to price ourselves out of the building market.

If I may take this opportunity, I would like to read portions of a report entitled *Export Market Possibilities for Railroad Ties* by Mr. Sadikin Djajapertjunda, the Director of Market for the Directorate General of Forestry of Indonesia.

Export market possibilities for railroad ties
By Sadikin Djajapertjunda

Efforts to increase the marketing of Indonesian forest products, primarily timber, have been initiated since 1971. These efforts have been directed towards the sale of products that had not found a market, as well as opening new market fronts. Based on preliminary investigations, one of the most potential products would be railroad sleepers.

The type of timber generally accepted on the market for sleepers are the sinker varieties like:

- Keruing/apitong (*dipterocarpus* species).
- Kempas (*koompassia* malaccensis species).
- Merbau (*intsia* bijuga species).
- Lara (*metrosideros* species).
- Balau/Damar Laut (*hopea* species), and others.

In this relation, it can be noted that the above varieties presently still have a very limited market, because the harvesting is more difficult and the sale price is not too attractive. Due to the above conditions, these woods have not been exploited on a large scale. In 1973 these varieties constituted less than 6% of the total Indonesian production.

According to our survey, these varieties still constitute about 10% of the total standing stock of Indonesian forests. Theoretically it would be possible to produce around two million cubic meters of round wood, which then represents some 10 to 12 million sleepers.

Some other reasons that have made the production of railroad sleepers less attractive are:

1. In spite of a relatively simple process, the production by conventional sawmills will definitely increase the cost.
2. Special transportation methods will be needed to bring the logs from the forests to the sawmill.

3. Loading equipment is still manual, thus making loading cumbersome and more expensive.
4. Ports that can handle the export of railroad ties are limited.
5. Sawmillers generally are not interested because of more difficult processing and lower profits.
6. The present offered price is relatively still less attractive than the direct sale of logs.

Some suggested solutions

1. Production costs have to be lowered.
2. A portable sawmill that can be transported from site to site would be preferable to a conventional sawmill.
3. Production has to be massive (large scale) to achieve low cost.
4. All transportation has to be systemised and well organised with the use of forklifts, etc., so skilled labour can be utilised to the highest efficiency.
5. The organisation and production has to be specialised (for sleepers).
6. A single collecting center would be advisable to expedite loading, export (and treatment).

Organisation

The organisation of this undertaking should be specialised and worked out separately. In this way costs can be further reduced. Therefore a specific company that solely produces and markets railroad sleepers should be set up.

The production of railroad sleepers will be according to the following principles:

1. The company will set up units for railroad ties production equipped with a portable saw unit. (An appropriate type would be the Wick Transportable Mill made in the United States.)
2. These units will be coordinated from one place. These coordinating centers will function as collecting, treatment, bundling, and loading centers for export.
3. The units can be spread out in such a fashion that logs can be easily found and sawn timber can be easily transported to the collecting point. The supply of logs can be arranged through a long-term work contract with H.P.H. holders, primarily for the supply of sinkers like keruing and kempas. Depending on the agreement, the logs have to be taken from the forest or delivered to the mill site. The company will pay the H.P.H. holders in the form of a fee if the timber is still standing stock, or as a domestic purchase if the logs are delivered to the mill.
4. The logs will then be sawn into the required sleepers and sent to the collecting points for selection, bundling, loading and export. For this transport, the units have to be provided with trucks, barges/lighters and tugboats. Possibly barges will be less expensive in case all the units are located along a river.
5. At the collecting points the following has to be provided:
   a. Unloading of feeder transport
   b. Grading
   c. Bundling equipment
   d. Loading equipment
   e. Other equipment

Size of the coordinating/collecting unit

If for instance one Wick Sawmill can produce 400 sleepers or about 30 cubic meters a day (theoretically one Wick saw has a capacity of 7 meters an hour), then to produce enough for one shipload of 10,000 meters, two months will be needed.

If one month is accounted for 25 working days, this will be 50 days for two months, so each Wick mill will produce 50 x 30 meters or 1,500 meters. For each single loading point therefore will be needed 7 to 8 units of sawmills to provide a load of 10,000 tons for each ship with a frequency of one loading every two months.

I am happy to say, that since this report was written, a suitable joint venture between an Indonesian and foreign company has been formed, a concentration yard site selected, and rapid progress is being made toward the realisation of this project.

The opportunities for utilising transportable sawmills in Southeast Asia are *vast*, and *now* is the time to get in on this rapidly expanding and profitable business.

Figure 8.1. This transportable sawmilling operation is located on the Batu Islands, off West Sumatra, Indonesia.

Figure 8.2. This mangrove swamp borders important forest area on the Batu Islands, Indonesia.

Figure 8.3. Logs are debarked by hand in the jungles of Indonesia.

Figure 8.4. Skid logs, placed under the transportable sawmill, enable the tractor to fill the skid ramp.

Figure 8.5. A hand winch is used for pulling the logs along the skids and under the mill for sawing.

Figure 8.6. A pier facility is necessary for the loading of timber onto ships.

Figure 8.7. Railroad ties are stored on the pier prior to being loaded onto the ship.

Portable Sawmills for Heavy, Hard-to-Reach Logs (Part II)

KLAUS H. SCHMIEDER
Managing Director, T-Export (Hamburg) GmbH
Hamburg, West Germany

Transportable sawmills do have a wide *field of operation.* Nearly every sawmiller will be confronted with one specific field of operation, which is waste-wood utilization. We all know that big quantities of commercial and noncommercial timber are wasted each year, due to the wasteful present system of logging in tropical rain forests.

Further investigations in Ghana, Nigeria, Brazil, Sarawak, and other countries have shown that for each cubic meter of logs extracted from the forest, nearly the same amount of wood exceeding 45 centimeters (18 inches) in diameter was left in the forest. This wood usually is of bad quality and low value and would therefore not pay for extraction, loading, and transport (see Figure 9.1). Therefore this waste-wood should be utilized directly on the spot; that is, in the forest, as waste is the most relevant factor in deciding on the rentability and feasibility of the logging operation.

Trials have proved that much waste-wood could be utilized economically by an integration of mobile sawmilling into road building, for example in:

- Tropical rain forests and swamp forests (for instance, the Alan forests of Sarawak, Sabah, and Kalimantan).
- Plain forests—for instance, in the Ashanti province of Ghana, in Beninstate (Nigeria), and Ivory Coast.
- Hill forests—for instance, on Borneo, Puto Range (Liberia), Elburz Mountains (Iran), and on Mindanao (Philippines).

This was the waste-wood utilization as one field of operation for transportable sawmills.

A second field of operation could be the conversion of logs to be felled for building roads through the forest. Traveling bandsaws, for example, were used in Iran for converting the trees that had to be felled for building the forest roads.

The trees felled on the roadline were skidded to the future working site of the bandsaw. Sawmill operations commenced well ahead of the road-building work. When the road finally reached the site of the mobile saw, all the logs had been sawn and the lumber stacked, so that the air-seasoned material could be hauled off by trucks. Otherwise this timber would have been lost for lack of motorable road connections. Now this timber has been utilized and sold in the domestic market, and a considerable proportion of the road-building costs was covered by the profit from the converted wood.

Using portable saws for timber production outside the forest

Apart from industrial sawmilling based on logging projects there are also very good chances to develop a kind of sawmilling industry based on the timber production outside the forest (see Figure 9.2). Semi-mobile sawmills, for instance, allow for commercial operation in this area. Despite more labour intensive operation these small saws could compete with big fully-mechanized mills for these reasons:

- *Better utilization of the raw material.* In different trials using small saws in several tropical countries, an out-turn between 60 and 80 percent was achieved, compared with 45 to 60 percent out-turn of industrial sawmills.
- *Lower costs for the raw material.* Portable saws could convert dimensions and qualities of round wood on an economic basis unsuitable for industrial conversion. Furthermore, defect logs of inferior quality can better be utilized on small saws than in fully mechanized plants.
- *Less transport costs.* Small saws are moved to the raw material, while the raw material has to be brought to the mill for mechanized conversion.
- *Less demand for capital investment.* The requirement of investment capital for small saws is considerably lower and could mostly be financed outside the capital market.
- *Less demand for technical and management know-how.* Training for the efficient operation of small saws could be achieved in a relatively short time, with domestic labour. Furthermore, practical operation of small sawmills seems to be ideal for training future management of sawmilling industries.

Types of mobile sawmills

There is a wide choice of mobile sawmills on the international market, and they differ considerably in technical and economical respects. Feasibility studies have been made on nearly all types of mobile sawmills, so I would now only like to mention their main features:

- *Gang saw mounted on trailers.* The heavy trailer can only be moved on good roads. The impact on the moving parts calls for a strong and really heavy construction of the vehicle. Installation of the gang saw as a mobile saw unit does not allow an automatic feed so that the high capacity can in reality not be utilized. Furthermore the diameter of the logs to be sawn is limited.
- *Bandsaw mounted on trailer.* This saw is more applicable in tropical rain forest operations, but it is very costly. Again the log feed isn't fully solved yet. Energy consumption is very high, rating at approximately 100 horsepower. The saw is mounted too high on the trailer so that all logs to be sawn have to be lifted up to this level.

- *Portable dimension saw.* For occasional conversion of big logs this equipment is the smallest and cheapest kind of mobile saw. It does not, however, show good economical results, due to the relatively high operation costs and low productivity. This is due to the transport; that is, bringing the saw from log to log and fixing it up.
- *Chain rip saw.* The operation and the sawing quality of this saw is too low and too much waste is involved. It is quite good for splitting big logs into workable dimensions (see Figure 9.3).
- *Mobile multiblade circular saw.* This saw is designed for the conversion of smaller logs. In tropical rain forest operations, it is used for converting small logs into lumber and splitting and edging flitches, made by horizontally moving bandsaws.

Semi-mobile log bandsaw

Various kinds of mobile and semi-mobile sawmills have been tested in various parts of the world for utilizing waste-wood and other timber. The semi-mobile, horizontally traveling bandsaws have been found to be the most economic tool for this operation as investment-costs of operation are low and the machine could be operated by low-skilled labour forces. This horizontal log bandsaw is a solid-built machine and has the advantage that it could nearly come completely assembled to the end-user. Final assembly is rather easy to achieve by utilizing an easy-to-make concrete foundation for stationary sawmilling. Although a semi-mobile unit, this machine is used in many cases as a stationary machine, due to its low investment-costs, creating medium sized industry at relatively low costs per working place.

The second foundation which can be made is an easy-to-make sleeper foundation which still gives a firm foundation for log and machine. Please note in Figure 9.4 that log and machine do have the necessary counter weight to give the final stability.

This log bandsaw can be loaded quite easily on timber trucks or on special saw transporters without dismantling. This means that the machine can be loaded directly from the rails onto the respective transporter.

Depending on whether or not voltage is available, the main drive can be done by electric motor, ranging only from 30 to 50 horsepower, by diesel engine, not exceeding 45 horsepower, or by a petrol power-unit, again not having more than approximately 45 horsepower.

Due to its easy operation and the centralized operating place, it does not require much skill to operate the saw from a standing platform. Operation can be done by domestic labour, after only a short time of training. High cutting accuracy achieved by a solid-built machine including strong saw blade guides guarantees fast cutting speeds and high cutting accuracy.

This is also achieved by a uniform chainfeed, operated through an infinitely variable gear drive. The semi-mobile log bandsaw is able to handle bigger logs up to 170 centimeters (68 inches) in diameter on the "through-cut system."

Depending on the quality and dimension of the raw material and the specification of the sawn products a good out-turn of between 60 to 80 percent is achieved by board production. Apart from board production, flitches can be edged or sleeper cants can be produced, but, with a considerable drop in productivity.

Productivity per man per shift is of course much below stationary sawmilling, since supplying logs to the saw is, in all mobile saws, labour intensive and cannot be mechanized. Stopping the machine due to the delay for replacing the logs can be avoided, however, on the semi-mobile bandsaw, if the rails are three or four times as long as the average log. While the first log will be converted, the second log will be placed in position and a third log is waiting in position to be converted in case of unforeseen delay in supplying the logs. Therefore the machine is in continuous operation. The logs are rolled into position either by domestic labour or by rope winches.

In comparison to other mobile-sawmills the productivity of the horizontally traveling bandsaw is quite satisfactory. The main advantage is the fact that on relatively low investment costs a very wide range of dimensions could be converted economically. Three men are needed to operate the semi-mobile log bandsaw which means one bandsaw operator and two men to remove the lumber after the cut.

In order to provide a continuous production flow you need more men, of course, for:

- Rolling the logs to the saw, depending on whether or not rope winches are used.
- Supervision, if required.
- Site preparation, if required.

The exact machinery and staff, however, can only be chosen by knowing the exact details on every single project, of course.

Saw blade sharpening is mostly done on simple sharpening machines, again being very simple in operation and having low investment costs (see Figure 9.5). These grinding machines are able to do a universal job and sharpen all tools used in this sort of sawmilling. These tools include:

- Bandsaw blades for the main mill.
- Circular saw blades for edgers or crosscut saws.
- Woodworking tools, such as planing knives, milling cutters, etc.

No production sharpening machines are necessary, however, because simple and strong machinery does a sufficient job in simple sawmilling.

Figure 9.1. Waste-wood of low value and bad quality can be utilized by an integration of mobile sawmilling and road-building.

Figure 9.2. Semi-mobile sawmills allow for commercial operations within the forest.

Figure 9.3. The chain rip saw is capable of splitting big logs into workable dimensions.

Figure 9.4. An easy-to-make sleeper foundation adds stability to a semi-mobile sawmill.

Figure 9.5. Saw blade sharpening is usually done on simple sharpening machines with low investment costs.

CHAPTER 10

Forest Products Utilization in Indonesia

MUHAMMAD SUNARYO HARDJODARSONO
Director, Forest Products Research
Bogor, Indonesia

Indonesia is endowed with vast forest resources. Of the roughly 120 million hectares of forest land, approximately 22 to 24 million hectares are economically accessible, and capable of producing 22 to 34 million cubic meters of logs.

These forest resources are managed on a decentralized basis. Up to the late sixties only the teak forests in Java and some pine and agathis plantations were managed regularly, but following the timber boom development since the early seventies, provincial forest services have been built up more effectively.

The utilization of forest resources has grown explosively since the enactment of the laws concerning the investment of foreign and domestic capital in Indonesia, respectively, in 1967 and 1968. Exports of logs has grown from about 500,000 cubic meters in 1967 to more than 18,000,-000 cubic meters in 1974.

In Indonesia the establishment of industries in general is considered a very important element in developing and stabilizing the economy, and the timber sector has also to give its fair share in this development. It is expected that ultimately 60 percent of total log production will be processed domestically.

During the first five year development period (PELITA I), 112 plants have been constructed or applied for, having a total intake capacity of 8,000,000 cubic meters. All of these plants will be operating by the end of PELITA.

Of the above total number of plants, 82 are sawmills comprising a total installed intake capacity of 4,700,000 cubic meters. The types of machinery and plant layout are mostly chosen for Malaysian prototypes not geared specifically to local conditions, and therefore they do not operate at maximum efficiency. Machines are of different makes, of which Malaysian, Taiwanese, Korean and Japanese are the most frequently used.

A lack of skills, primarily skilled operators and experienced foremen in sawing and maintenance, saw doctors, and mill-managers, has resulted in the production of substandard lumber. Nevertheless some headway has been made in starting an export market, which is presently growing rapidly, that is, from 34,095 cubic meters (1970) to 337,948 cubic meters (1973) in four years. It is hoped that this growth will continue after the world economy has recovered from the present recession.

The domestic demand is also growing rapidly owing to the successes obtained by the PELITA's. Presently projected relevant economic targets will make a vigorous growth of domestic demand for sawn wood on the order of 10 to 12 percent per annum, most likely.

There are many problems remaining to be solved, but both international and domestic market outlooks give ample reason to be optimistic for further growth of sawmills, in particular, as well as other forms of wood processing in the future. Tables 10.1-10.7 give relevant data.

Timber resources

Indonesia has generally been considered a lucky country with regard to its possession of natural resources in the form of timber, among other resources.

Official estimates of forest lands in Indonesia range from 90 to 120 million hectares, depending on one's view regarding whether the some 30 million hectares of lalang grass and secondary growth created by ages of shifting cultivation can be considered as belonging to the forest area or not (Table 10.1). Included in these 90 to 120 million hectares are one million hectares of plantations consisting of hardwood stands, primarily teak (650,000 hectares), mahogany, rosewood, etc., and some 200,000 hectares of pine stands, here and there being intermingled with a few thousand hectares of agathis plantations. Increased efforts had been made in the past to expand these scattered plantations of conifers during the first five year development plan (PELITA I).

During the current second five year development period (PELITA II) this expansion program is being more than doubled.

The remaining forests consist of various types of natural stands, depending on their habitat: tidal forests lining the coastal land of eastern Sumatra, Kalimantan, and West Irian; swamp forests covering the coastal areas of eastern Sumatra, central Kalimantan, and southern Irian; the lowland Diptericarp forest which covers the plains in Sumatra and Kalimantan; or finally the highland forest which grows on the mountain slopes of nearly all major islands. Before the war, only about 14 percent of the total forest area was formally registered as forest reserve, about half of which with some form of demarcation. Only the teak forests in Java were managed on a regular basis.

Since 1967, a very extensive reconnaissance survey has been carried out, and approximately 54 million hectares of the forest area have now been covered by the so-called "orientation survey" using a very minimum sampling intensity (0.01 to 0.02 percent). This reconnaissance, however low in percentage, has given further insight into the economical potential of the timber resources. A very preliminary estimate indiates that about 22 to 24 million hectares have practical economic accessibility. Commercial volume increments of these natural stands vary from 0.5 to 2.5 cubic meters per hectare per annum on a selective logging basis according to the existing cutting regulations. Several locations have substantially higher standing stock. However, these are only exceptions and are very limited in extent. Based on the above data, the actual national log production of these forests can range from 22 to 34 million cubic meters per annum on a sustained yield basis.

The presence of industries can undoubtedly make a higher yield per hectare possible, since part of the waste usually left behind as unsuitable material for the export market can go to the mills and be processed into lumber, converted into back veneer or core material, or chipped into raw material for the board and paper industries. The extra quantity to be gained in this manner is theoretically substantial, although actual figures are not available.

As mentioned above, the increment of commercial volume of the existing natural forest is relatively low and the government has therefore adopted the policy of encouraging the establishment of the so-called "integrated" industries and of chipmills capable of processing wood and selling chips on an all-species-mixed basis. If this can be realised, then the practice of clearcutting followed by artificial regeneration can be promoted. In manmade plantations the annual increment of the stands will be increased six to ten times or perhaps more, depending on the species and site-class. This practice of clearcutting and artificial regeneration will eventually be imperatively necessary if one considers the projected wood demand and supply in the year 2000, when the world will have to house 6.9 billion people, with Indonesia contributing at least 240 million to that number (Table 10.7).

With those projections in view, the government in Indonesia has multiplied its reforestation budget sixfold in this second five year development plan. And investigations are in progress in what ways the reclamation of the more than 350 million hectares of lalang grass and young regrowth can be made feasible for private initiative, either for forest crops or any other form of agriculture. This program will be closely tied to efforts to "resettle" the population practicing shifting cultivation, a task which is neither easy nor small.

All the forest land in Indonesia is, by law, state domain. The management of forest resources has, however, been decentralised since 1957, although the major decision policy is still in the hands of the central government. Not all provinces had an adequately equipped and organised Regional Forest Service at that time. The booming log market since 1969, however, not only necessitated the provincial governments to develop a proper apparatus to manage their forest resources, but also gave them the means to do so. Considerable progress has been achieved in resource management during PELITA I. Many problems are, however, still to be solved.

Utilization of forests

Considering the extent of resources available, the utilization of forest resources up to the late sixties was insignificant (Table 10.2), but a sharp increase was noticeable after the enactment of Law Number 1, 1967, and Number 6, 1968, which regulated the facilities to be granted to foreign as well as domestic investments and Law Number 5, 1967, concerning the basic matters in forestry.

The early seventies were marked by a further rapid growth of forest exploitation. Production rose well over 22 million cubic meters, while export topped 19 million cubic meters in 1973 (Table 10.2). Somewhat lower figures were realised for 1974 because of the recession taking place in almost all consumption centers of tropical hardwoods. I believe that 1975 will show a further decline in production and exports although there is opinion that the market will recover, although slowly, during the second half of the year. It is expected that production and export will settle on a somewhat lower level than the 1972 figures. Further

growth of log production will be "normal" and not as spectacular as what was taking place during the early seventies. In this growth the domestic market is expected to increasingly play a stronger role, if related economic targets of PELITA II are realised.

Wood-based industries

As in other countries, in Indonesia the role of industry is considered vital to economic development and economic stability. During the five year development period, the timber sector developed quickly and proved itself to belong to the first five major foreign currency earners in the country along with mining (oil), plantations, and agriculture. It is therefore expected that during the second five year development period, the timber sector will continue to grow and wood processing will certainly help alleviate social-economic pressures within Indonesia.

On the general economic platform the timber industry is expected (a) to continue to sustain and, through processing, increase its contribution to the country's foreign currency earnings through value added from the produce exported; and (b) create more job opportunities either by direct employment or by multiplier effect.

In addition to that, from forestry's own point of view, the establishment of processing facilities will improve the rendability of the stands by its capacity to absorb some of the waste now left behind in the forest. Another more subtle effect is that the ownership of processing facilities will exert a positive influence on logging companies to better observe the sustained-yield principle in their operations. The concept that in one phase wood processing industry will form one of the conditions to logging is already taken up in the logging contracts. At the same time it is recognized that some of the log production will always have to be exported as logs, for economical as well as political or other reasons. It is therefore stipulated in these contracts that eventually (ten years after commencement of the logging operation) 60 percent of the total production be processed domestically. The remaining 40 percent of log production will continue to be available for international trade.

Deviation from this pattern may of course occur, as dictated by general or local conditions. It may be that at one time the establishment of processing facilities at one location is not (yet) feasible, thus calling for log export higher than 40 percent. On the other hand, conditions could be such that 100 percent log processing at one time at one location for one species is very desirable and attractive, thus creating a decrease in logs for the export market.

As has been mentioned above, the ultimately desired form of processing is the "integrated" one. This will combine flexibility of operations with the production of least waste and best diversification of produce.

Companies are for those reasons persuaded to consider the establishment of combinations of processing from the start. In several situations where the existing supply of raw materials warrants, companies are urged to aim at attaining the highest form of utilization and go all the way to paper projects, or at least pulp production. Three locations are presently committed to sustain a pulp and paper mill, although actual conditions show that several elements are still inadequate to do the job, including raw material supply as well as harbor facilities and roads, still to be improved.

In many locations, veneer and plywood manufacturing will be the highest form of processing possible accompanied by sawmills as a form of integration. The completion of this combination with a hardboard or particle board mill to utilize the waste, will be rarer. Some complexes will try to attain waste utilization by setting up a chipmill, i.e., at locations where bulk transportation is possible.

In most locations sawmilling will be the only form of processing possible and may be completed with moulding facilities and joineries. One or two companies are starting to produce prefabricated house components, and development in this direction is likely to continue. The government expects housing industry to grow about 9.0 percent per annum. Since the building of medium- and low-cost housing is having priority status, this is likely to create a demand-pull for sawmills in general and prefab components in particular. Many of the timber concessions presently operating are in their second five years of operations and are trying to establish industries. The present situation of wood industries development is illustrated in Table 10.3. It is expected that projects presently under construction or in the stage of submittance of application will be already operating by the end of PELITA II (1978).

The plants registered in Table 10.3 are all newly established plants since 1969 or were at least "modernized" during that period. Of the total of 112 plants, 82 (73 percent) are sawmills accounting for 4,700 million cubic meters (50 percent) of the raw material demand.

There are 13 (12 percent) integrated mills demanding 25 percent of the raw material intake and 15 (14 percent) plywood mills accounting for 17 percent of the intake of logs into the plants.

Sawmilling industry in Indonesia

In the last ten years a considerable progress has been made in sawmilling equipment and technology in Indonesia. The impact was notably felt from 1970 onward when logging concessionaires, especially outside Java, brought with them modern equipment for the conversion of their logs.

Pit-sawing, for example, which was the most common method of log breakdown during the period 1940-1950, was gradually replaced between 1950 and 1969 by a simple mechanical method. Circular saws powered by truck or jeep engine were quite popular during that period for sawmilling, producing about 5 to 25 cubic meters of sawn timber per day on a one-shift basis. The number of sawmills which existed before 1969 was reportedly 943 with a total installed capacity of around 4,095,000 cubic meters annual intake. Most of these mills were relatively old. After 1969, 41 more mills were added so that the total number of sawmills presently in existence in Indonesia is 984.

During the period 1974-1975, 47 more sawmills have been recommended and are ready to start operation sometime in 1976. In 1976 the number of sawmills in Indonesia will thus reach a figure of 1,025 with a total rated input of 8,816,000 cubic meters per year, the detail of which can be seen in Table 10.4.

Gangsaws were only employed in some sawmills in Balikpapan, Banjarmasin and Sampit (Kalimantan) but they were considered as inefficient or uneconomical because of the conditions prevailing at that time in regard to skill, spare parts supply, etc.

Bandsaws are most preferred at present, particularly those types with annual capacity of up to 18,000 cubic meters or roughly about 20 to 30 cubic meters output per shift per day. Some sawmilling companies in fact employ bandsaws of 60 to 80 cubic meters or sometimes 100 cubic meters daily output. In regard to the production and consumption of timber in Indonesia, a rapid increase was observed until 1973. This trend is expected to continue for some time in the future considering a rising demand for sawn timber at home as well as outside Indonesia. The anticipated or projected consumption of sawn timber and wood raw material for each fifth year between 1970 and 1990 can be seen in Table 10.5.

The following are some characteristics of sawmill development in Indonesia:

1. During early existence, some sawmills received only work orders from outside and they thus acted as service mills. After 1969 some mills became independent and operated irrespective of orders from outside.
2. Seventy-five percent of the present sawmills have designs and layouts similar to those existing in Malaysia and Singapore.
3. Each resaw unit is independently powered and the power used is usually small, for example the power required to operate a 48-inch band resaw seldom exceeds 40 horsepower.
4. Construction of most of the sawmills was not based on a well planned theoretical design and technical layout; production flow and manpower efficiency were not given proper consideration. These designs rest wholly in the hands of the suppliers or sellers of the machines without taking into consideration the local conditions.

Cutting saws most commonly found in Indonesia are the circular saws and the bandsaws. The former is of the simplest type, i.e., single saw blade. The circular saw is directly driven by its own motor or via a V-belt connected to a central driving shaft. The log is simply mounted to a log carriage which in turn is pushed by hand.

Most of the bandsaws brought into Indonesia came from Taiwan, Japan and Korea. These machines (especially Taiwan-made) are preferred by the local sawmillers because of their low prices. These sawmillers have been using various saw types, from portable circular saws to semiautomatic or fully automatic bandsaws. Most of the bandsaws are of the vertical type, and they often have specifications as follows:

1. Pulley diameter varying between 1100 and 1200 millimeters.
2. Average maximum cutting height 1200 millimeters.
3. Saw blade speeds between 30 and 40 meters per second.
4. Return speed of carriage exceeding 150 meters per minute.
5. Feed speed up to 150 meters per minute.
6. Power requirement between 30 and 80 kilowatts.
7. Blade thickness between 1.1 and 2.0 millimeters.
8. Blade width between 100 and 140 millimeters.

A relatively "modern" saw mill in Indonesia may have the following layout of machineries:

1. One unit of 60-inch breakdown bandsaw with auto feed carriage.
2. One unit of 44-inch breakdown bandsaw with feed carriage.
3. Five units of resaw table employing 42-inch bandsaw.
4. Semi or fully automatic log transportation of life roller.

In some Japanese-designed sawmills, crosscut saws are included in the layout, while in most sawmills using design from Taiwan, Malaysia or Singapore, the crosscut saw is replaced by chainsaw for the crosscutting of logs in the log pond.

As mentioned above, most of the sawmill owners in Indonesia entrust the design of their plants to the suppliers of equipment from overseas. As a consequence the layout and composition of the machines are still doubtful or inefficient. In some cases, for example, the set-up and type of the machines are not suitable for handling long and large sized logs or the feeding and cutting speeds are inappropriate. In some sawmills there has been a practice to resaw timber on a table bandsaw using a short "guiding bar" without using a carriage. This habit has resulted in an uneven saw edge as the control depends only on human touch. Another factor which had been responsible for the production of low quality sawn timber in the past and at present is insufficient saw doctoring facilities in most of the sawmills. As most of the sawmilling equipment is of overseas origin—for example, Sheng Feng and Tong Yang from Taiwan; Sam Yong from Korea; Fuji, Iseda Tanaka, Akimoku, and Tsutsui from Japan; and Canali from Italy—it is not surprising that repairs and procurement of spare parts have been some of the bottlenecks to be overcome. Dealers who can readily supply the necessary sawmill spare parts or maintain workshops for repair are not yet sufficiently available. Hence many sawmills must set up their own workshops for repairs and maintain an unduly amount of spares inventory. For many mills this proves to be a heavy financial burden.

Sufficient spare parts to be made readily available is indeed a "must" or otherwise production will be stagnated. Saw doctoring is also a problem.

If saw doctoring workshops could be established in big cities, the service of which would be made available to sawmillers, this would indeed be good news for these people. Perhaps this possibility must be given proper attention as this will not only cause a possible lowering of the capital required for investment by the sawmillers but will also introduce saw doctoring technique which is still lacking in many sawmilling operations in Indonesia. It should be noted here that the most common mistakes in saw doctoring operation in Indonesia are wrong or improper filing of saw teeth and the use of unsuitable tooth patterns.

Markets

The figures earlier shown in Table 10.2 give a picture of how the export market exploded. This upsurge is even more staggering if we consider that in 1964 the total timber exports were less than 100,000 cubic meters. This explo-sion was the result of a coincidence of several supporting factors, such as:

1. The general business boom that took place at the consumption centers of the world. Processing capacities grew and caused a rise in demand in Japan, Korea and Taiwan for wood products and raw material.
2. The change of economical climate that took place in Indonesia since 1966, making it more attractive for foreign (and domestic) private capital to come to Indonesia.
3. The turn towards more conservative policies in the traditionally log-exporting countries like Malaysia and the Philippines calling for a decrease of logs exported.

All these factors together have created the rush for timber in Indonesia.

Looking at Table 10.2, please note also that the balance of logs not being exported is also growing during the period indicated by the table. This difference between production and exports was 1.6 million cubic meters in 1969, and grew to 6.8 million cubic meters in 1973. This means that the domestic demand growth is matching that of exports.

This domestic demand for logs parallels the growth in industrial capacity shown in Table 10.3. A substantial part of the produce is being consumed domestically. A growing percentage, however, is also finding its way to export market as can be seen from Table 10.2, which shows that the amount of exported sawn timber rose from a mere 34,000 cubic meters in 1969 to about 338,000 cubic meters; more than tenfold in four years.

Conclusion

1. In Indonesia there are still vast untapped timber resources, of which about 24 million hectares are economically accessible and capable of producing up to 34 million cubic meters of logs annually under present conditions.
2. The resources' management, which 10 years ago was very minimal, is undergoing improvement. Investigations are meanwhile in progress to put unproductive areas, devastated by past shifting cultivation practices, again into production, preferably with private initiative, although the government to play a ground-breaking and pioneering role. Reforestation and rehabilitation funds have been more than doubled for the second PELITA as compared to the first.
3. In the utilization of forest resources in Indonesia, industrialization is expected to play a major role in increasing the foreign currency earning of the country and in providing more opportunities for employment. It will also improve the per-hectare rendability of the forest resources and exchange forest management practices on a sustained or perpetual-yield basis.
4. PELITA I has seen the establishment of some processing facilities. New additional plants are in construction and a few more applications are coming in. There are 112 new plants which will be in operation by the end of PELITA II in 1978, with a

total designed capacity of more than eight million cubic meters of raw material intake per annum.

5. Of these plants 82 or 73 percent are sawmills and this number will continue to increase. The first mills to come in are mostly of simple layout after a fixed pattern not adapted to specific local or market conditions. Younger generation mills are showing improvement in the layout system. As concession operators become more informed about markets, market standards in scaling and grading, and development of sawing technology, better equipment and a better system will be adopted.

6. Although domestic processing of logs was and is being promoted in Indonesia, it is envisaged that part of the production, plus or minus 40 percent, will still be available for international trade.

7. The economical and social conditions in Indonesia are still developing. There are still many things to be thought out and many problems to be solved. But the government is endeavouring to attain vigorous industrial growth and it is hoped that many presently occurring obstacles will be cleared for industrial development in general and for wood industries in particular.

8. The export market is presently experiencing a slump for logs. It is even more difficult for processed timber. We are however confident that this will finally improve. The domestic market is coming on strong and can be expected to sustain this trend if the development programs in Indonesia have any success at all.

9. I may venture to conclude that the wood processing industry has made its first few steps in Indonesia and further developments will have a fair chance to succeed.

Table 10.1. Indonesia timber resources (x 1,000 Ha)

Island	Total land area	Forest area	Percent
Java and Madura	13,200	2,99	22,6
Sumatera	47,400	28,42	59,9
Kalimantan	53,900	41,47	76,9
Sulawesi	18,900	9,91	52,4
Nusa Tenggara	7,400	1,48	20,0
West Irian and the Moluccas.	49,600	37,50	92,1
Total :	190,400	121,770	67,1

Table 10.2. Production and export from Indonesia.

Year	Production (logs) (m3)	Export	
		logs (m3)	Sawn timber (m3)
1969	5.299.000	3.705.000	–
1970	10.899.000	7.350.000	34.095
1971	13.706.000	10.761.000	80.759
1972	17.717.000	13.891.000	132.195
1973	26.297.000	19.488.000	337.948

Table 10.3. Development and spreading of wood industries in Indonesia

No.	Province	Type of Ind.	Stage of Development	Approx Installed capacity 1.000 m3 intake	Potential AAC in Region 1.000 m3	Estimated invest. US $ 1000
1.	A c e h	1 it	P	32.5		
		3 sm	C + P	137,0		
		2 sm	A	80.0		
		1 pw	P	35.0		
		6 pl		334.5	800	14.000
2.	N. Sumatra	1 pw	C	80.0		
		1 pw	P	69.0		
		2 pl		149.0	450	7.800
3.	W. Sumatra	1 sm	A	115.0	350	2.200
4.	R i a u	2 it	A	160.0		
		4 sm	P	108.0		
		2 sm	A	84.0		
		1 pw	C	50.0		
		9 pl		302.0	1.200	22.900
5.	Jambi	2 sm	C	160.0		
		2 sm	A	220.0		
		1 pw	A	120.0		
		5 pl		500.0	850	18.400
6.	S. Sumatra	1 sm	P	48		
		4 sm	A	160.		
		5 pl		208	480.0	3.830
7.	Lampung	3 sm	P	180.0		
		1 sm	A	160.0		
		4 pl		240.0	200	5.400
	Total Sumatra	3 it / 25 sm / 5 pw	33 pl A+C+P		4.330	74.530
8.	W. Kalimantan	3 it	P	690		
		13 sm	P	590		
		1 sm	C	36		
		3 sm	A	220		
		2 pw	A	120		
		22 pl		1.656	1.350	30.880
9.	S. Kalimantan	1 it	A	90		
		2 sm	P	78		
		1 sm	C	72		
		1 pw	P	34		
		5 pl		274	460	10.650
10.	C. Kalimantan	5 sm	P	440		
		4 sm	C	210		
		5 sm	A	620		
		3 pw	C	435		
		1 pw	A	120		
		18 pl		1.825	4.400	68.050

Table 10.3   (continued)

No.	Province	Type of Ind.	Stage of Development	Approx Installed capacity 1.000 m3 intake	Potential AAC in Region 1.000 m3	Estimated investm. US $ 1000
11.	E. Kalimantan	2 it	P	220		
		4 it	A	790		
		5 sm	P	216		
		10 sm	A	600		
		1 pw	P	75		
		1 pw	A	175		
		2 ch	A	12		
		25 pl		2.068	11.500	98.960
	Total Kalimantan	10 it) 49 sm)70 pl 9 pw) 2 ch)	A + P A+C+P A+C+P A		17.710	208.540
12.	S. Sulawesi	1 sm	P	48		
		1 sm	A	6		
		1 pw	P	50		
		3 pl		104	240	2.560
13.	C. Sulawesi	3 sm	A	84	380	2.860
14.	M a l u k u	3 sm	P	108	1.200	1.000
	Total Sulawesi + Maluku	8 sm) 1 pw) 9 pl	A+P		1.820	6.520
		112 pl	± 8.000		± 23.860	± 286.590

Notes :   P = in production;   C = under construction

A = application submitted

it = integrated industry ; sm = sawmill

pw = plywood will ; ch = chip mill

pl = plants

Table 10.4. Development and spreading of sawmills in Indonesia

LOCATIONS	Concession / Non Concession	PRODUCTION				CONSTRUCTION				RECOMMENDED				TOTAL			
		No.of Enterprise	Intake m3 (x1000)	Output m3 (x1000)	Investment US $ (x1000)	No.of Enterprise	Intake m3 (x1000)	Output m3 (x1000)	Investment US $ (x1000)	No.of Enterprise	Intake m3 (x1000)	Output m3 (x1000)	Investment US $ (x1000)	No.of Enterprise	Intake m3 (x1000)	Output m3 (x1000)	Investment US $ (x1000)
SUMATERA	Concession	12	530,8	275,5	11747,5	1	100,0	60,0	1204,7	12	732,8	360,4	15833,4	25	1363,6	695,9	28785,6
	Non Conces.	431	999,6	465,1	-	-	-	-	-	-	-	-	-	431	999,6	465,1	-
J A W A	Concession	4	-	-	-	-	-	-	-	-	377,0	-	2815	4	-	-	-
	Non Conces.	262	556,9	274,0	-	-	-	-	-	-	-	-	-	262	556,9	274,0	-
KALIMANTAN	Concession	25	1339,6	645,5	18408,7	6	219,8	189,9	7436,5	18	1461,1	1002,1	42325,5	49	3110,5	645,5	68170,7
	Non Conces.	159	2107,5	1022,1	-	-	-	-	-	-	-	-	-	159	2107,5	1022,1	-
SULAWESI	Concession	1	48,0	-	-	-	-	-	-	4	90,0	30,0	2714,5	5	138,0	30,0	2714,5
	Non Conces.	70	367,1	182,9	-	-	-	-	-	-	-	-	-	70	367,1	182,9	-
NUSATENGGARA	Concession	-	-	-	-	-	-	-	-	-	-	-	-	-	-	-	-
	Non Conces.	7	6,9	3,5	-	-	-	-	-	-	-	-	-	7	6,9	3,5	-
MALUKU	Concession	3	108,0	20,0	1000,0	-	-	-	-	-	-	-	-	3	108,0	20,0	1000,0
	Non Conces.	8	23,4	12,0	-	-	-	-	-	-	-	-	-	8	23,4	12,0	-
IRIAN JAYA	Concession	-	-	-	-	-	-	-	-	-	-	-	-	-	-	-	-
	Non Conces.	6	34,5	17,3	-	-	-	-	-	-	-	-	-	6	34,5	17,3	-
INDONESIA'S TOTAL	Concession	45	2026,4	941,0	31205,2	7	319,8	249,9	8641,2	34	2283,9	1392,5	63688,4	86	4720,1	1391,4	103534,8
	Non Conces.	943	4095,9	1975,9	-	-	-	-	-	-	-	-	-	943	4095,9	1975,9	-
	TOTAL	984	6122,3	2916,9	31205,2	7	319,8	249,9	8641,2	34	2273,9	1392,5	63.688,4	1025	8816,0	3367,3	103534,8

Table 10.5. Estimated consumption of wood products in Indonesia.

Classification	1970		1975		1980		1985		1990	
	Total (1,000m3)	Per capita (m3)	Total (1,000m3)	Per capita (m3)	Total (1,000m3)	Per capita (m3)	Total (1,000m3)	Per capita (m3)	Total (1,000m3)	Per capita (m3)
A. INDUSTRIAL WOOD :	10.290	0.083	11.536	0.085	13.429	0.087	15.741	0.091	18.533	0.094
a. Logs	4.086	0.034	4.570	0.034	5.287	0.034	6.102	0.035	7.077	0.035
b. Conversions	4.963	0.041	5.712	0.042	6.637	0.043	7.782	0.044	9.765	0.045
c. Plywood	182	0.002	328	0.002	386	0.002	531	0.003	713	0.004
d. Pulp Wood	726	0.006	952	0.007	1.235	0.003	1.592	0.009	2.037	0.010
B. FUEL WOOD :	86.919	0.718	83.600	0.613	78.413	0.500	71.277	0.403	61.009	0.300

Sources: 1) Timber Trend and Prospects in the Asia Pasific Region 1961, FAO.
2) Wood World Trend and Prospects 1967, FAO.
3) Yearbook of Forest Products 1968. FAO.

Table 10.6. Inter-insular imports of wood in Indonesia (1000 cubic meters)

Location	Year	
	1973	1974
Jakarta	598	720
W. Java	17	23
C. Java	23	55
E. Java	62	130
Other Regions	175	231
Total	875	1,159

Table 10.7. Projections of domestic consumption of woods.

Year	Population in million	Consumption per capita (m3)	Total Consumption (Million m3)
1974	127,6	0,083	10,4
1975	130,6	0,085	11,1
1976	133,7	0,094	12,6
1977	136,8	0,103	14,7
1978	139,9	0,113	16,4
2000	245,0	0,50 (0,90)	123

CHAPTER 11

Best Opening Face

H. C. "CARL" MASON
President, H. C. Mason & Associates, Inc.
Gladstone, Oregon, USA

The term *best opening face* is, I think, quite confusing, and accordingly, I would like to explain why this is so. As far back as I can remember, the term *opening face* has been used to describe the opening of a saw log which, in most cases, I am going to refer to as a *grade* log.

It did not make any difference whether it was a hardwood log or a softwood grade log—the cutting pattern was still much the same. The technique, if applied properly, was generally to open an opening face that in most cases could be defined as the minimum opening face. In the United States, due to our size standards, the minimum opening face was as close to four inches as possible, and the technique or rule for finding the best opening face, or opening face in the absence of knowing specifically what was in the log, was to open to the point where you could find the first usable commercial size wood in the log, which was generally four inches. In sawmills where production was not a particular criterion, but where recovery was a criterion, this is generally what happened.

The next step in sawing was normally what we call the four quarter jacket board or a one-inch board and that generally was maybe one or more one-inch boards. The purpose was to develop the face wider and see what it looked like in terms of grade. One-inch boards were used because in the high grades in flat sawn lumber, generally speaking, a one-inch board was as good a commercial product as you could get, and if you went deeper than that, the losses were rather substantial.

Let me explain that. If you look at a slab, there is a rule of thumb that says if this angle exceeds 45 degrees, it should have been cut thinner. So if you have a very fine slab, a very thin one, and this angle was always less than 45 degrees, keep it thin. That just means that if you do not do it, you will find that you have excessive loss in the edging that is thrown away. It is strictly a matter of geometry. But in further definition, this is what we always meant by opening face, and it was the *best* opening face.

The converse of this is what usually has been happening over the last 20 or 30 years in sawmilling, primarily because capital investment, the cost of a mill, is a large part of sawmilling costs. Historically, labor was one of the next largest costs. In the United States, when saw logs were $5 on the stump, labor was still $10 or $15 and capital costs were $10 or $15 per thousand. And today, when saw logs are $300 on the stump, our costs are still pretty much the same.

In the years past, in order to improve production, to get more wood through the mill, the sawing pattern generally was thick-slice. The method was to step in and take a big slab off and get into the meat of the wood as soon as possible. Usually that slab got sent to a resaw. I can say almost categorically that doing this the sawyer would never find a minimum opening grade or best opening grade, because the rule would say, or just the law of probability would tell you, that sawing this way and coming out where you should is nearly impossible, especially since this was an arbitrary line that was established with no rules whatsoever. Therefore, if you did saw in this way, most of the time you would lose the board. This is a fundamental lesson in the sawing of a grade log.

My company, H. C. Mason Associates, had been diagramming logs by hand on a drawing board for years using templates, trying to find how we could get more usable lumber, more salable lumber, out of a truncated cone log. In 1968, we learned to do this arithmetically. By a formula, we learned to use a computer to diagram a log and find the best sawing pattern.

About 1970, the Forest Products Laboratory in Madison, Wisconsin, under the direction of Hiram Hallock and Dave Lewis, began to use a method similar to ours. We had been trading information for years, and they came up with a program and we came up with a program.

Our program we called "Random Dimension," because that term described it to us. When the Forest Products Laboratory produced their program and had it operational, they needed a name for it. They used the name "Best Opening Face." In volumes one and two of *Modern Sawmill Techniques* published by Miller Freeman Publications, there are articles on "Best Opening Face" by Hiram Hallock and Dave Lewis. The articles say that the program was primarily designed for small logs, generally under 20 inches in diameter, or that the only products to be considered were dimension lumber—one-inch and inch-and-a-half boards. In doing so, you take a log and consider it in two ways—half-taper and, conversely, full-taper. Either way. And you would consider sawing the log just two ways. One is with a center-cant in the middle and the other is live sawing such as you would do with a frame saw where you just slice it up alive.

You can do either with B.O.F., the Best Opening Face program. It is an arithmetical program very similar to our Random Dimension program, in that if run on a computer under any circumstances, for any market and using any conversion cost index, you will find the best sawing pattern for a true truncated cone with a known small end diameter, known large end diameter, and a known length.

Once you have that program and once you know what is in the log, then the problem is deciding how to implement it, and that is really the key to today's sawmill technology. We have to start somewhere, and the easiest place to start is to go back to this. Say, "Okay, if we know where all these lines are in here—it doesn't make any difference which side—we can identify it. We can put a line here. We can say, 'This is it, Mr. Sawyer, whoever you are, whatever equipment you are using, if you will accomplish this on your first line and then if you will saw all this pattern all the way through, you will get optimum conversion out of a log.' "

QUESTION: I think I understand what Mr. Mason is speaking of in regard to this programming of cutting. In light of the need to cut selectively to avoid or try to minimize defects which exist, I think it would make it rather more difficult to implement this program in relationship to tropical hardwoods, and I think most of us would be pleased to hear Mr. Mason comment.

ANSWER: Let me explain. The Best Opening Face program and our Random Dimension program were never designed for green cutting. They were designed for regrowth timber generally under 14 inches in diameter, producing nothing but one-inch boards and dimension frame lumber generally in the United States. The Best Opening Face as a technique, as developed by the Forest Products Laboratory, has very little application to old growth cutting logs; whether it is hardwood or Ponderosa pine or spruce or fir or whatever, the results of the problem are the same.

CHAPTER 12

Mill Layout for Maximum Efficiency

H. A. "HARRY" HALLEWELL
Manager-Equipment Sales, MacMillan Jardine (M) Sdn. Bhd.
Kuala Lumpur, Malaysia

Conditions around us are changing at an ever accelerating rate. In order to meet these changing conditions sawmillers of today must plan seriously about the requirements for a mill which will meet future demands. No one can predict exactly what future requirements will be, but we must utilize our past experiences, analyze present trends, and plan to the best of our ability what we want to build for the future. We must utilize logs to produce the best line of products to suit the mill design and intended market.

Many jobs today in Southeast Asia are being done by hand—barking, heavy log handling, transferring of lumber, removing waste from the mill, etc. As time passes, it will become and is becoming more difficult to interest people in this type of work. In conjunction with the decline in availability of these people, there is a continuing increase in the cost of labour. It may only be in the initial stages in some of Southeast Asia but the cost of labour is sure to accelerate in the future.

My general feelings regarding mill layout, which includes the equipment selection and placement thereof, is that we should be designing with the objective of maximizing the productivity of each individual machine, which in turn will yield the maximum production with the minimum number of machines. This design theory does suggest some degree of automation but not for the purpose of reducing manpower. The phase of straight manpower reduction through automation will come when the conditions warrant it and our designs of today should make provision for future installations without the need for high additional capital cost.

Initial considerations

A number of basic questions must be answered prior to deciding the basic flow design or layout which will suit your needs best. Some of these are:

1. Basic wood species.
2. Log consumption plan.
3. Log quality.
4. Log size (diameter and length distribution).
5. Outline of sizes you wish to cut and amounts of each (marketing).
6. Degree of flexibility you wish to have in your mill.

Mill layout and equipment selection go hand in hand. In selecting the equipment, one must size the basic components considering the production capabilities and characteristics of each machine and then attempt to blend them into an overall mill layout.

I would like to consider the following categories or areas in discussing mill layout:

1. Log debarking and handling.
2. Primary breakdown.
3. Secondary breakdown.
4. Final or remanufacturing.
5. Trimming.
6. Lumber stacking and handling.
7. Waste removal.
8. Saw maintenance.
9. Mill maintenance.

Let us look briefly at each one of these areas. I shall attempt to outline some basic flow patterns plus considerations in selecting the particular piece of equipment. The procedures and comments which I will outline are intended only to help in assessing the various aspects of mill flow patterns and operations. The comments do not cover all aspects of the operation and are merely intended to stimulate thoughts which should be beneficial in obtaining the best plan for your conditions.

Log handling and debarking

In the majority of mills in Southeast Asia today, debarking is done by hand. In future mills, mechanical log debarking equipment will become more common. If you do not wish to install this type of equipment today, at least plan to leave the necessary area clear where the equipment could go in the future.

Two basic types of debarkers are in use (Figure 12.1):

1. Rosser head type.
2. Mechanical ring type.

A simple straight-through flow or the "C" shaped flow-through is as efficient as any. The selection of the debarker or debarkers will depend on your production requirements and log size distribution. A Rosser head type will probably suffice up to a production of *100* tons per shift (*60 MFBM*).

In conjunction with debarking you will need some type of log cut-off saw. Again, there are two basic types:

1. Chain type.
2. Circular saw.

Production is limited on the chain type but one can always double up on the number of saws. The circular saw type is much faster; however, the saws are difficult to handle and they do require more saw maintenance. A good thought to keep in mind about debarkers is to leave some means of getting logs into the mill without passing through the debarker. They have been known to break down on occasion.

Primary breakdown

The most common and probably the best type of

breakdown is the band headrig and carriage combination. Some considerations for this key production area are:

1. Reliable log decks—this does not mean long storage decks. Make maximum use of space and move logs efficiently.
2. Operator should be in a position to see all operations in his area.
3. Have responsive log turning and log handling equipment. North American mills have made extensive use of air log turners and loaders. I feel that some of the Asian designs for log turning (chain type) offer better and less violent control.
4. Carriage should have good, positive, fast dogging and should have air cushions on the carriage knees along with automatic setworks.
5. Carriage drive should be responsive and capable of keeping the amount of time the saw is not cutting to an absolute minimum. This will yield maximum breakdown at the headrig rather than just cutting for subsequent manufacturing operations.
6. Have an efficient lumber take-away system with a clear, smooth straight edge adjacent to the carriage outfeed rollcase.

A bandmill is capable of cutting a great deal of lumber if that is what it is doing, namely "cutting." Too often you see carriages taking excessive time to set, return, get up to cutting speed, loading the log, turning the log slowly, etc.

If properly maintained, the bandmill/carriage combination provides extreme versatility and produces straight accurate lumber. The piece is held well under control and the sawyer always has the opportunity to see the log or piece before he decides where the next cut will be (Figure 12.2). If your production requirements are such that you can do the majority of your log breakdown on one carriage and bandsaw, the addition of a secondary deck on the outfeed is advantageous. This deck can be used to deposit large cants and in turn these cants may be transferred back onto the carriage for further processing. If in the future you wish to increase your production facilities you can simply add another bandsaw/carriage combination using this cant storage deck as the feed to the second headrig.

I would like to make a few points regarding bandmill selection. A large number of sawmillers today are talking about high strain both with conventional and air type strains. The two basic benefits of high strain are:

1. Reduced kerf.
2. Improved sawing accuracy.

These two criteria must go hand in hand. Reducing kerf with a subsequent reduction in accuracy buys you nothing.

There are a number of specific things you should consider when selecting any bandmill. These are:

1. Wheels should be cast of a high strength alloy with high hardness factor (i.e., 220 Brinell or higher).
2. Wheels should be dynamically balanced.
3. Only the highest quality bearings should be used.
4. The strain system should be designed to yield high speed response and good damping. Strain points should be of high quality, precision manufactured, and easily replaceable.

5. The bandsaw should be equipped with a replaceable type pressure guide system.

There is one other thing to keep in mind and that is that the most successful sawing mills, whether high strain or conventional, are those with highly skilled saw doctors. Unfortunately, these people are not easily found.

Secondary breakdown

After the material is cut on the main headrig, it can now be directed to several different machines:

1. Another bandsaw/carriage combination.
2. Single or multiple bandmills.
3. Edger.

The selection of these machines will depend largely on the type of logs you are cutting and the variety of sizes you wish to make. If the wood contains a lot of defects, then single or multiple bandsaw is preferable and yields the best recovery. If your logs are free from major defects, the use of an edger will tend to maximise your production with a reduced amount of capital outlay.

If you select a second bandsaw/carriage combination, versatility, accurate sawing and good recovery will result.

With twin and/or single bandsaws, the latest arrangement has been to use a stationary line bar and have the bandmills set on a sliding base. With precision type setworks the bandmills can be set to within thousandths of an inch and the line bar remains rigidly stable as it does not have to move. With a twin or single band resaw and recycle system, an efficient arrangement is shown in Figure 12.3.

The operator is standing where he can see the operation well enough and still is in a position to handle or help smaller pieces if they have to be stood on edge. The recycle systems can be either manual or automatic. Present labour rates cannot really justify the highly automated system. Regardless of this, the system must provide for continuous cutting to maximise the productivity.

I believe that edgers also have their place in hardwood manufacturing, depending on where and how you use them. In the last five or six years, there has been considerable advancement made in sawing technology. The large flat friction pads with some type of lubrication used as saw guides are the heart of the system. Many types of materials and configurations have been used that have enabled kerfs to be reduced from the old 11/32-inch to 3/8-inch range down to .100 with accuracy of ± 1/32-inch. These guide systems have been used on conventional shifting saw edgers as well as single and double arbor circular gangs (to replace sash gangs).

Final or remanufacturing

Final manufacturing or remanufacturing is basically confined to the use of resaws and/or small edgers for purposes of sawing previously cut larger pieces, edgings, etc., to a final size or remanufacturing those pieces which may not meet specifications because of grade, length, size, etc.

Resaw systems which utilize quick return devices or are arranged so that pieces can be returned quickly to the infeed by manual means are very effective. Again, as mentioned several times previously, the idea is to keep the machine cutting lumber, not "air." This will yield maximum productivity on a unit machine basis.

Trimmings

Two basic types of trimmers are in use today:

1. Multiple saw selective type.
2. The two-saw opposing trimmer (two saws arranged on opposite sides of a common trim table as in Figure 12.4).

For productions which are not excessively high (in order of *100* tons per 8 hours or even greater), I prefer to use the two-saw type. You cannot trim out defects unless they are on the end of the piece (at least not easily) but you do not tend to overtrim. A man has to work to overtrim, and if he gets tired by the end of the day he will not overtrim. The one disadvantage of this type of trimmer is that some retrimming is usually required at a later date.

With the selective trimmer it is very easy to "hack" wood up. It is a very effective, versatile machine but must be watched closely to see that wood is not being wasted. Too many times have I seen an operator get bogged down and just put all the saws down and run lumber through just to get rid of it.

Stacking and sorting

I will not tend to dwell on the sorting aspect because certainly automatic sorters are a long way away. I always feel if the lumber gets out of the mill, one can always find some way to sort it and stack it for kilns, air drying, or subsequent processing. Just make sure you have adequate room for a proper sorting chain. They always seem to be too short!

Waste removal

Since some 25 percent to 35 percent of the volume of wood coming into the mill is going out in the form of waste, the waste removal system should not be taken lightly. If the funds permit, build the mill well off the ground and leave room for waste conveyors below. As fibre becomes more and more valuable for future needs, conveyor systems with chipping and/or refuse devices will become an integral part of the mill operation. Low pressure blowing systems can be used effectively for sawdust removal. Conveying systems (and there are many types) do not need to be elaborate but a few key conveyors can assist greatly in keeping the mill producing efficiently.

Saw maintenance

Successful saw maintenance or saw doctoring is one of the most important factors in determining the success of a sawmill operation. The most advantageous location for the filing room is above the mill operating floor. Band and/or circular saws can be changed in the minimum amount of time and with the minimum of handling. If the filing room cannot be placed above the mill it should be located in a central location close to the key operating machines. Wherever the filing room is placed, beware of vibration, particularly if you are using carbide equipment. This is one of the problems of having the filing room above the mill floor.

Choose your saw filing equipment carefully and plan carefully the layout of the filing room itself.

Maintenance

When designing and laying out a mill do not forget about the maintenance aspect. Maintenance is an integral part of the process, not one of those necessary evils one must put up with. Try to standardize on as many items as you can to eliminate the variability and necessity to keep large and costly volumes of spare parts. Make sure that maintenance shops are centrally located relative to all mill facilities. Designing for minimum and ease of maintenance in mind will pay good dividends in the long run and will tend to simplify any preventative maintenance programing.

Typical mill flow

To combine some of the foregoing principles into a typical layout for a potential hardwood mill, I would like you to consider briefly the layout in Figure 12.5. This layout, as mentioned initially, is meant basically to spur some questions in your mind if you are contemplating either revising or building a new mill. It is certainly not meant to be "the" answer.

Summary

To summarize I would like you to consider some following general comments when designing a mill:

1. Some mill equipment does not have to be installed in the initial stages but—it is a good policy to try to leave space to put it in later.
2. Select machinery which is capable of doing what you want and which will provide long and efficient operation. Consider the use of mechanical equipment where heavy labour would normally be required. Functions affecting mill output and accuracy should be done by mechanical devices.
3. Ensure that lumber flows systematically from one machine to the next. This will tend to reinforce the maximum productivity principle (Figure 12.6).
4. Do not forget about the safety aspect in your mill. As more sophisticated machinery and techniques are used, hazards to health and safety are increased. Take care to guard all conveyors and moving parts of machines.
5. Consider the noise factor. There will be a time in the future that regulations regarding noise levels will be enacted. Consider them today, do not get caught in the future.
6. Your people are your key to success. Treat them well, be concerned about their well-being and job environment and give them the necessary training to do their jobs properly.

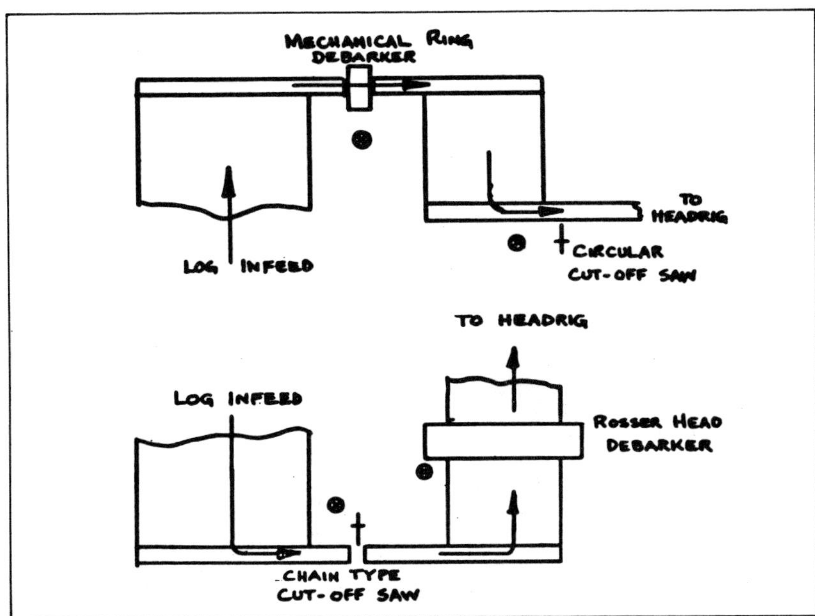

Figure 12.1. These are the two basic types of debarkers.

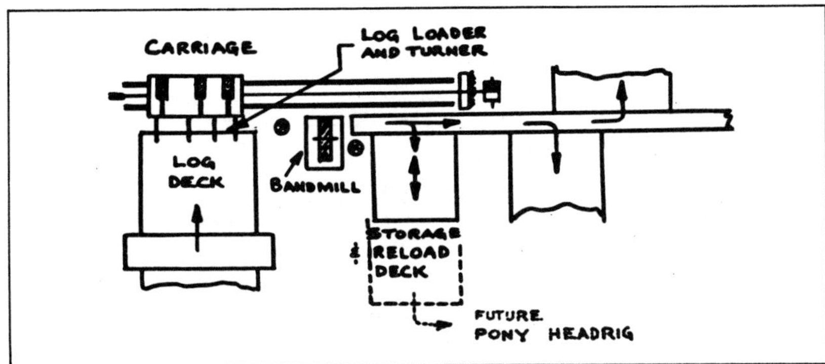

Figure 12.2. A properly maintained bandmill/carriage combination provides versatility and produces straight accurate lumber.

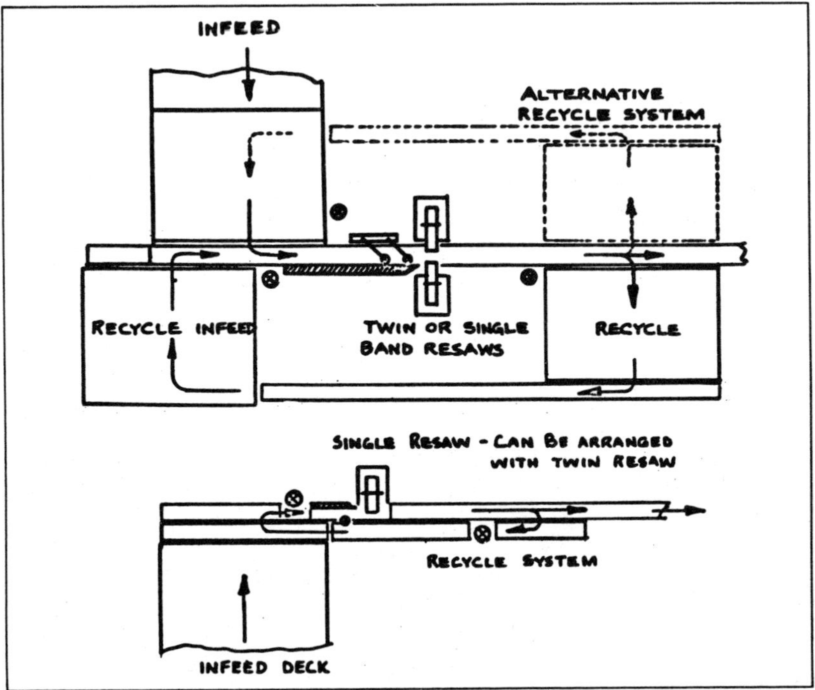

Figure 12.3. With a twin or single band resaw and recycle system, an efficient arrangement is important.

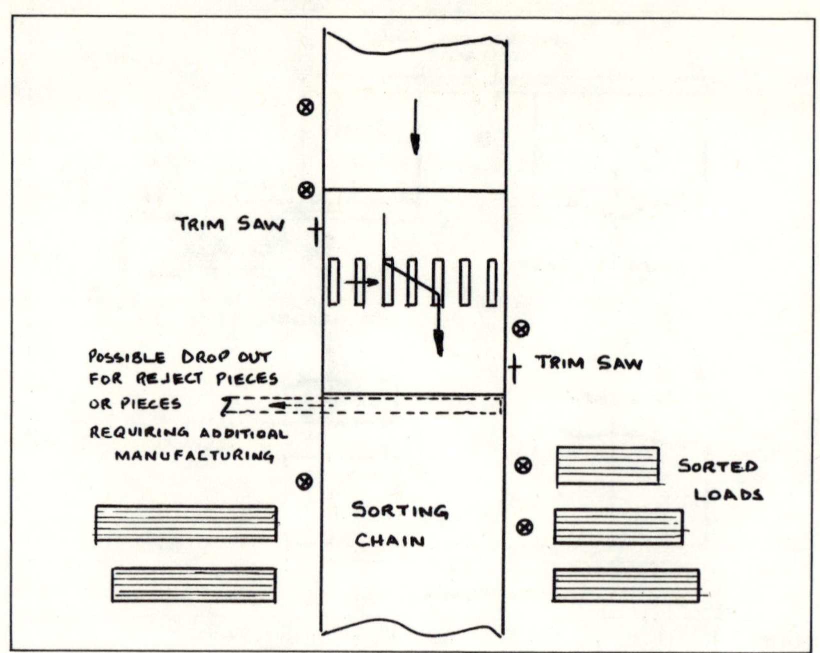

Figure 12.4. The two-saw opposing trimmer works efficiently for productions which are not excessively high.

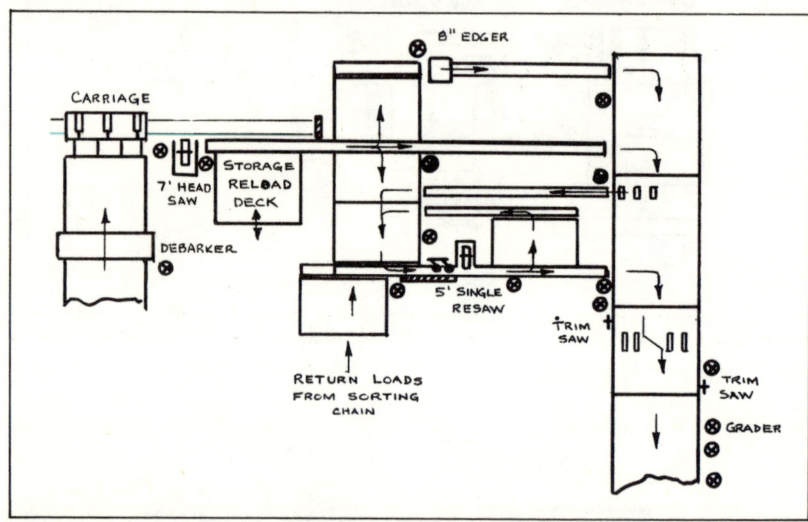

Figure 12.5. A typical layout for a hardwood mill.

Figure 12.6. Lumber should flow systematically from one machine to the next for maximum productivity.

CHAPTER 13

Canadian Sawmill Equipment to Increase Accuracy

PHILIP KENDRICK
Vice President-General Manager, Canmillex Sawmill Exports Ltd.
West Vancouver, British Columbia, Canada

At the outset it should be said that perhaps the title of this chapter would be more appropriate if it was "Canadian Sawmill Equipment to Increase Accuracy and Production." The two conditions are obviously intertwined as will be evident from the following remarks.

Traditionally, the largest percentage of sawn lumber produced in Canadian sawmills is destined for overseas markets. These markets, very competitive, have been responsible for the development of closer tolerances of the lumber to be sold. At the same time, higher production levels were sought to satisfy the ever-increasing world demand. Obviously when sawmills were demanding a higher production, the need to increase efficiency through accuracy in sawing became the main consideration. Accuracy of cut, therefore, has, in the past as well as in the present, been the target of sawmill machinery design engineers.

Accurate sawing carries many other benefits for the sawmiller. Lumber recovery from the log can be increased due to less culling and cutting closer to required dimensions, thus effectively reducing the planing depth.

The cost of maintenance in a sawmill has also been found to increase noticeably when machinery is used which is not required to produce to fine tolerances. It is well known that a piece of machinery that is not performing well can generally be a real problem to maintain. We should also note at this point, of course, that where precision sawmilling is practised, and maintenance is generally of a protective nature, increased mill output is very noticeable through a minimum of downtime.

Components of the sawmill

We now direct our attention to the components of a sawmill where accuracy is sought. These resolve themselves to production machines, feeding devices to machines, and the equipment for the conveying of finished, semi-finished and waste products, and it is at this point that some may doubt the importance of a reliable waste system. It has been proven many times that a backed-up waste system can impede mill flow and quality of the finished product.

In modern Canadian sawmills, such consideration has been given to the removal of waste material with the emphasis on creating a clean environment for the production equipment. Sawdust and trimmings which remain in the vicinity of the product being sawn can quickly upset the settings of machines and affect the performance of the saws. We are all aware of the problems, for instance, of a saw without sufficient gullet area to clear the sawdust.

In order to facilitate waste removal, Canadian mills are generally designed with sufficient space between the mill floor and ground floor to allow for an efficient transfer

from one waste conveyor to another. Steep slopes can then be made in the transfer chutes to prevent material hang up.

Production machines have been equipped with wipers and scrapers while adequate openings in the mill floor allow an automatic disposal of waste. Careful attention has also been paid to conveyor mechanisms to prevent jamming or plugging which indirectly can result in mill inaccuracies as well as undesirable down time.

Although sawmill chip production may not yet be a major operating factor for Southeast Asia sawmills, it is nevertheless of much importance in nearly all sawmills in Canada. As chips represent a major source of revenue, much attention has been placed on the production of accurately cut chips with a minimum of overs, fines and sawdust. It is interesting here to make mention of the fact that poor knife maintenance on chippers not only reduces the production of dimensionally stable chips and increases the quantity of unacceptable fines, but also drastically increases the power requirement for chipping and the mill costs for supplying this power.

The equipment to be supplied for close tolerance lumber cutting actually commences with a debarker. It has been proven that debarked logs provide an opportunity for much better log scanning and the consequent lumber recovery, as well as the opportunity to cut more accurately. Rocks and other foreign objects which are often present in the bark of logs can be the cause of saw snaking as well as saw damage which results in poor cutting. It has been noted that by eliminating the down time caused by rock and other foreign material, an increase of production in the magnitude of 2 percent has been achieved.

For the purpose of this paper we will consider a sawmill using a carriage and bandmill type headrig arrangement rather than a frame saw. The carriage and bandmill headrig configuration is notoriously known for its capability of sawing for grade, great dimensional variety, and, last but not least, the accuracy so much desired. The combination of these three virtues, if we may use this terminology, will undoubtedly lead to a good market position and increased revenues.

The first production machine to be discussed, therefore, is the sawmill carriage. In recent years, designs have been produced which permit very rapid knee adjustment and maintain log alignment in relation to the saw. Many years ago rack and pinion as well as screw block designs were common for driving carriage knees while the rugged chain drive, however, has become prevalent in recent years. It is relatively easy to take up any possible chain stretch, whereas lost motion through wear on racks or screw blocks was a cause of inaccuracy which cannot be readily overcome. Continuous accuracy in the Canadian designs is assured through the use of readily replaceable wearstrips whereby close attention is given to the mode of lubrication to assure a smooth operation. The modern tong dog design

---

This paper was delivered at the Seminar by Adrian J. Kuypers, Sales Engineer, Canmillex Sawmill Exports, Ltd.

has shown great advantages over the standard board and dog arrangement which has been well known in the past. This is due to the ability of the tong dog to provide a positive pull back, assuring that the log or cants are tightly retained against each knee during all phases of sawing. The bottom dogging feature which is common with the tong dog further assures stability of the log or flitch.

In addition, mention should be made of a wide range of options which are available and are designed to assist the headrig operator in turning the cants, adjust for taper, lessen the impact of a bouncing log on the carriage frame, and select the feedspeed to suit the variety of species sawn. Carriages are presently designed to accept optional features as they become desirable to suit the nature of the production (Figure 13.1).

The carriage setworks and controls presently used in Canadian mills provide such accurate adjustment that the element of human error in knee setting has been eliminated. The most sophisticated controls provide for backstand setting, thus freeing the operator of having to calculate the number of boards in any given cant. Setworks designs have been much improved in recent years and this could be the subject of a paper in itself. Suffice it to say that with moderate attention from normally skilled maintenance people, there is no cause for setworks accuracy to be reduced.

In recent years, innovations in bandmill designs have produced marked improvement in cutting. However, we must assume, of course, that saw doctoring is of good quality as this is the prime prerequisite of any bandmill operation. Sawmill operators tend to underestimate the importance of tooth profiling and saw tensioning, a complaint often voiced by sawfilers. The advent of the highly sensitive strain has produced a marked improvement in lumber quality particularly when high feed rates are required. This feature also appears important when cutting large heavy logs as saw stability is much improved.

The indiscriminate use of the term "high strain" has often brought about a misunderstanding with regard to particular applications. High strain, per se, should be used in context with response times, at such time a saw enters the log to prevent snaking. Therefore, the term "highly sensitive strain" should be considered when speaking of this phenomenon. The need for such sensitive strain devices may not be required in many applications. Several bandmills, including Canmillex equipment, provide an optional use of this feature. The devices now commonly used combine the normal weighted strain arrangements with a quick acting air or hydraulic device which overcomes the inertia of the weight system. This means that saw response to unusual resistance is reduced to a fraction of the time that was hitherto experienced. In addition, the use of properly designed pressure guides and wheel scrapers assists in a correct traveling of the saw and virtually eliminates saw snaking.

A few additional comments regarding the constructional aspects of the bandmills are in order. In anticipation of both low and high production levels, while processing small and large diameter logs, it is of extreme importance to consider the effects of vibration caused in the bandmill frames. It is for this reason that most Canadian sawmills use the double column design with the wheels mounted in between (Figure 13.2). The top wheel arbour, then, is preferred to be of the non-rotating type with bearings mounted in the hub. This eliminates the flexing of the otherwise rotating shaft. Candian bandmill requirements, in common with all machinery, tend to be of a very rugged design which results in inherent machinery stability and a long life with low maintenance expectancy (Figure 13.3).

The edgers play an important role in Canadian sawmills and are designed and manufactured to produce large quantities of accurate dimension lumber. We, in Canada, manufacture a large variety of models to suit many requirements and are often asked for recommendations which are welcomed due to the numerous applications, not to forget economic considerations. Edger frames are normally line bored, stress relieved and provided with spigot mounted bearings. The arbours are chrome plated with accurately cut keyways while in the case of larger machines, we often supply a splined key to increase saw stability. The most recently designed top arbour edgers provide the ability for a small amount of float on the collars and eliminates the possibility of wedger cutting (Figure 13.4). The saw guides, of course, must be benched and jigged for accuracy in relation to one another. Attention has been given to the removal of sawdust and to this end the over arbour edger is a much cleaner operating unit than the standard bottom arbour machine. The elimination of sawdust in the area of guides and collars is a worthwhile benefit in achieving accuracy. It is of interest also that the top arbour design provides for a cleaner cut through the edger, thus reducing the margin necessary for planing, resulting in increased recovery.

For the different settings of the saws, various shift arrangements are available from the hand-operated and semi-automatic to the preselect units. High production requirements in Canadian mills have sped up the development of these shifting setworks. The subsequent sophistication required prompted Canmillex to delegate this to our electronics division resulting in the air electronic setworks which is capable of basic one-inch increments as well as fractions of an inch by simply actuating a combination of push buttons on the control console. This, combined with adjustable rod ends and the accurate saw guides noted hitherto, results in lumber produced to closer tolerances.

The same redesigns as noted for the headrig bandmills have been applied to resaw bandmills. The standard roller feedworks of earlier years is now more popularly supplanted by an infeed linebar which supports the timber throughout its length and increases the straightness of the lumber cut. It has been found that if a linebar is not of a rugged design then lameness can occur in its operational length and this can affect setting accuracies. The linebar setworks as for the edger is normally of the cylinder bank design.

Canadian sawmills have recently much increased the productive capacity of resaws and we now produce twin and quadruple band machines. In these instances, saw shifting is achieved by complete frame adjustment on very long guides and this has been found to be both rapid and capable of producing a very accurate setting. This machinery requires hydraulic cylinders for motion control and a very high degree of positioning is achieved by the use of Dynaset. The Dynaset, which would require detailed discussion, is a hydraulic positioner controlled and actuated by a combined air and electronic arrangement which can be programmed for any setting combination to a fraction of an inch.

Basic mill construction

No discussion on the cutting of accurate timber would be complete without consideration of basic mill construction. The foundations must be properly designed and provide stability and alignment from one machine to another or to its conveying systems. The alignment of the log carriage to a frame saw which is particularly susceptible to foundation loading is a case in point. Likewise, shock absorption throughout the sawmill can be important in retaining machine accuracies. Timber construction is particularly good from this point of view and where steel frame building construction and machinery support is used, attention should be paid to shock absorbing pads under the footings of major production machines.

In conclusion we should make note of the importance of the operational personnel of any sawmill. Attention can be given to first class design and machines, but unless there is incentive and a desire by the mill operators to do a good job, inaccuracies for one reason or another will creep into the finished product. A key man in the Canadian sawmills is the saw filer who has great responsibility and is rewarded accordingly. The head sawyer and saw filer are by far the most important people in the sawmill and their performance can affect the attitudes of other mill personnel. Combine this with a planned program of preventive maintenance which reduces frustration due to unscheduled breakdowns, and we can achieve a degree of harmony of operation that will do much to augment capabilities of sawmill machinery.

There is much yet to be learned in North America if we are to attain greater accuracy and lumber recovery. The engineering departments of my own organization are constantly redesigning and striving to improve the inventory of new machines to achieve these objectives.

Figure 13.1. Carriages are designed to accept optional features as they become appropriate to production.

Figure 13.2. The double column design for bandmill frames minimizes the effects of vibration.

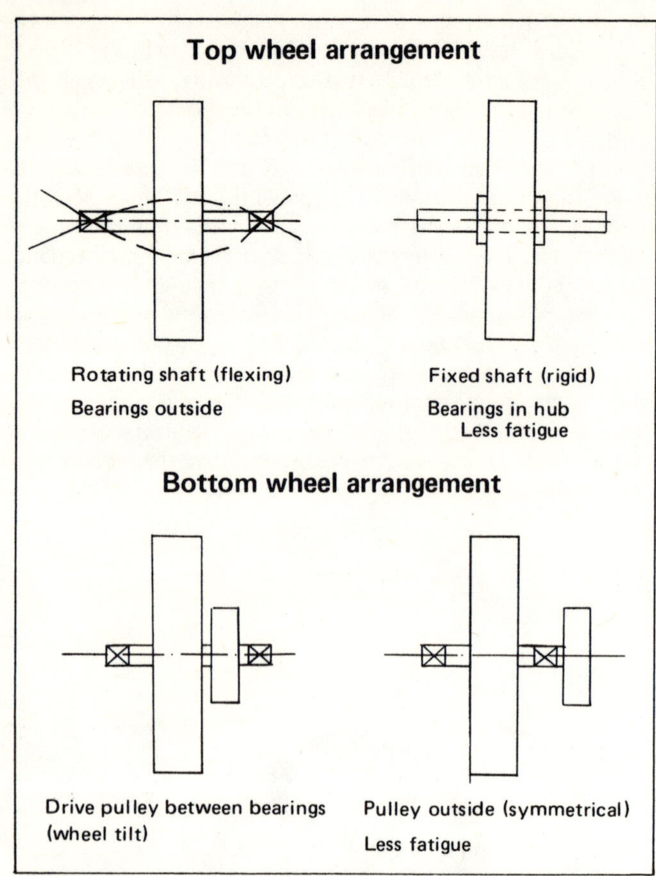

Figure 13.3. Top wheel and bottom wheel arrangements of bandmill.

Figure 13.4. This edger is designed to produce large quantities of accurate dimension lumber.

CHAPTER 14

Application of Modern Technology in Lumber Processing

ROBERT ARCIAGA
Export Sales, Canadian Car (Pacific)
Makati, Rizal, Philippines

Recent worldwide developments in the wood industry have emphasized the need for Southeast Asia to *diversify* its wood products and markets. Moving towards this diversification, it is inevitable that the region will adopt modern methods of production, equipment maintenance, and product marketing, in order to achieve the independent stance vitally needed to meet the effects of periodic fluctuations in market condition.

Southeast Asia's wood industry has reached a firm position from which to launch a "take-off" period towards modernization. Logging, and to some extent, salvage saw-milling, over the past 20 years have built up sound financial equity, experience, and performance records among local wood companies—factors which would readily attract foreign venture capital when or if this is needed.

Also, the development of an advanced plywood industry has produced a core of Southeast Asian technicians with high levels of expertise, so much so that the entry of modern sawmilling and other wood-processing equipment should present no problem where application and maintenance are concerned.

For this chapter it could be worthwhile to concentrate on two related topics:

1. Recovery of wood chips from sawmill waste.
2. Some recent developments in high-production machines.

Possibilities for chip recovery

The progress of pulp and paper manufacturing in the Philippines using indigenous tropical species indicates that wood chips will have a domestic market within the next few years, in addition to the growing requirements for chipboard, flakeboard, and waferboard.

In addition, the growth of broadleaved softwood tree plantations (also in the Philippines) with thinnings available three to four years after planting, and full harvest possible five to six years afterwards, in turn, indicates that the pulp and paper industry will be assured of more uniform, and perhaps better quality, chips for industry development.

In most of the existing sawmills in Southeast Asia, the bulk of mill waste is burned simply because no further use could be found. In cases where a wood-working and finishing line has been set up as a recovery facility, the waste generated by the finishing line itself has far exceeded pre-operational expectations. Additionally, in most cases, the finishing line also proved inadequate to the amount of wood waste produced by the sawmill, even though a number of mills have started using waste as boiler feed.

Let us take, as an example, the case of a sawmill with an input of between 4,500 to 5,000 cubic meters of logs per month, doing mostly salvage operations on logs not suitable for export and/or plywood grades. Lumber recov-

ery is assumed at 50 percent, or an approximate average of one million board feet a month. Mill waste also comes to an equivalent of one million board feet per month.

If 75 percent of the mill waste is converted to chips, and an additional 500 cubic meters of logs not suitable for sawmilling are included for chipping, the value added could be arrived at as follows:

$$500 \text{ cu.m.} = 17,650 \text{ cu.ft.} + 750,000 \text{ bd. ft. of mill waste}$$

$$= 80,150 \text{ cu.ft. total less } 5\% \text{ allowance for "fines" and other losses}$$

$$= 76,150 \text{ cu.ft. of chips available, in solid wood content}$$

1 selling unit of chips = 200 cu.ft.

1 unit of 200 cu.ft. chips = 72 cu.ft. solid wood content

$$\frac{76,150}{72} = 1.055 \text{ units available}$$

1.055 x \$30/unit x 12 months
= *\$380,000* value added/year of operation

A number of other factors, such as utilization or transportation and marketing, would of course have to be considered. However, to all intents and purposes the additional investment on equipment, storage facilities and handling machinery could very well pay back in one year or 18 months.

Selecting chip recovery machinery

The recovery of chips from mill waste at existing mills could of course be readily done with the addition of a roundlog and wastewood chipper, with chipscreen.

The selection of a roundlog and wastewood chipper would seem to be a straightforward exercise. However, since chipping knife upkeep and frequency of change constitute a major operating expense, it is necessary to have complete coordination with the manufacturer to determine basic operational requirements, such as the horsepower needed to drive the disc, the diameter and power needs of the disc itself, the configuration and number of knives most ideal for the wood species, chip size, and production rates involved. Type of feed, whether horizontal or drop, seems to influence the amount of "fines" produced in chipping hardwood logs. However, the added stress on both the knives and the disc by "force-feeding" through gravity could affect the frequency of knife change.

At any rate, the following features would have definite advantages for a chipper:

1. Ease of knife-changing.
2. Easy and accurate adjustment of knives and anvils.
3. Swept-back or non-radial knives will reduce machine vibration and chip variation, for improved chip quality.
4. Knife slots should be designed for free discharge of chips, to reduce chip damage.
5. Balanced wear plates will provide a smooth surface between knives.
6. An optional rechip spout should be available for small pieces and oversize chips.
7. Extra large spout can be provided for whole logs, slabs, and multi-wood pieces, with minimum "dead-space" in the spout area.
8. Shear bolt arbor design should allow disc to move back when a foreign object enters the spout.
9. Rigid chipper frame, preferably of stress-relieved, fabricated construction.

High production machinery

For new sawmills still to be installed there are high-production machines now available, designed for higher lumber recovery, more uniform chip production, and reduced fines and sawdust formation. For example, the Chip-N-Saw sawmilling and chip-recovery concept as applied in North America and Europe, could be applicable, with some revisions, in Southeast Asia for the following types of operations:

1. Clear-cutting of forest lands for disposition to agro-industrial purposes, or conversion to tree farm areas.
2. Recovery of chips and lumber from broad-leaved softwood harvests in tree plantations.

The basic unit in a Chip-N-Saw line is a "linear processing machine" in which processing of debarked logs to fully-manufactured lumber can be achieved in one pass. Up to 85 percent of the wood waste is converted to marketable chips.

To put it simply, the machine operates as follows: The log enters a chain infeed section, where it is centered and driven through the machine. The bottom chipping head makes the first cut, producing a "guide tongue," or spline, at bottom of the log. This guide tongue enters a matching profiled guide bar which continues through the machine, and controls the position of the log during the further chipping and sawing process. The second cut is made by two side chipping heads, producing flat reference faces on both sides. The third cut is executed by the top chipping head, which produces a flat reference face on the top of the log.

Having been processed on all sides by the chipping section, the cant is then presented for further manufacture by the saw section. Up to this point, no sawdust has been produced, and all waste has been converted to chips. The saw section (which could either be circular saws mounted on double arbors, or a quad bandsaw arrangement with high-strain) then reduces the cant to the required dimension lumber.

Chip-N-Saw application covers a log diameter range of 4 inches (10 centimeters) to 28 inches (70 centimeters) at butt, in lengths of 8 feet to 24 feet. In case of logs in the larger diameter range (20 inches to 28 inches), band resaws and edge chippers would be needed further downstream in the mill, to increase width recovery and obtain lumber of variable thicknesses.

Assuming a feed speed of 100 feet per minute (30 meters per minute) and a continuous supply of unsorted, debarked logs in the diameter range for which the machine is designed, production capability would be approximately 2,000 logs per 8-hour shift, of 16-foot average length logs, with recovery as much as 61 percent lumber, 33 percent chips, and 6 percent sawdust.

For application in Southeast Asia, where weak or defective log centers could be a problem in addition to log tension-release considerations, the ratio of lumber and chip production will vary to a certain extent. Also, it could be that only the chipping and canting operation should be incorporated as a "single-pass" section, as illustrated in Figure 14.1.

The completed cant could then be remanufactured through a combination of band resaws, edge-chipper, and perhaps a circular or reciprocating gang, to obtain the best recovery given log condition and lumber size requirements.

The Chip-N-Saw chipping heads are provided with extremely accurate setworks (hydraulic stacked cylinder or D.C. drive infinite-position ball screw) which enable the operator to handle various diameters of unsorted logs.

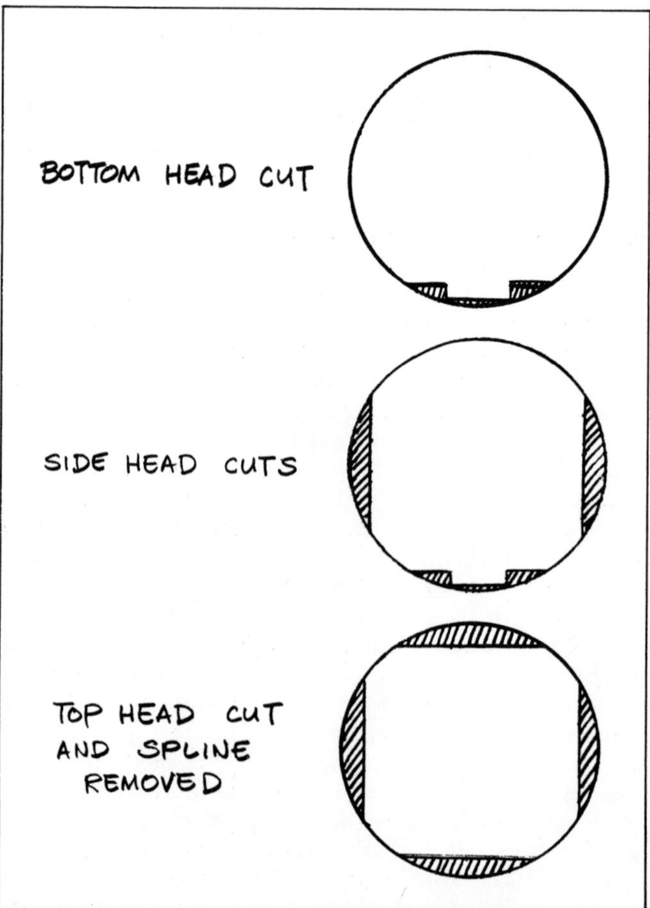

Figure 14.1. The chipping and canting operation should be incorporated as a "single-pass" section.

Other options are available, such as pre-select setting of chipping heads, automatic offset feeding, and even computerized gauging and setting, for optimum recovery.

Power requirements for the basic Chip-N-Saw unit ranges from 500 to 1,200 horsepower, depending on model, wood species, feed rates, etc. Peak power draw is normally around 60 percent. Moving into the processing of the large diameter hardwood trees, from 30 inches (80 centimeters) upwards, we now come to the recent development of the Chip-N-Saw vertical carriage.

The vertical carriage is an innovation to the conventional carriage in that the holding arrangement for logs is done through top and bottom dogging, through the use of pneumatic cylinders. This makes possible the multiple sawing of logs at each pass at the twin or quad band headrig, and provides the option of slab-chipping on the first pass to produce the flat reference faces on both sides (Figure 14.2).

The log can be turned at the carriage, and passed repeatedly through the multi-position twin or quad band headrig. The smallest center cant that can be produced is 4 inches (10 centimeters), for further resaw or gang-sawing.

The multiple-sawing of live logs by dogging at the ends is not a new innovation, but holding problems have, in the past, been encountered. I believe that by the Chip-N-Saw system, these problems can be overcome especially for the larger diameter pieces, because the vertical carriage has a number of dogging cylinders, top and bottom, running along the entire length of the log, with the pneumatic cylinders following the contour to retain firm hold (Figure 14.3).

The ability to position the log centerline at any angle in the vertical carriage, with generous leeway for off-set feeding, would make possible an additional 2 to 3 percent recovery. An example would be the feeding of "bowed" logs with the centerline at 45 percent angle immediately on the first pass. This saves the trouble of considerable slabbing on one side before a bowed log can be positioned adequately on a conventional carriage.

Figure 14.4 shows the log at 45 degree angle on the vertical carriage, with the effect of reducing the log "bow" as presented to the bandmill saw lines.

The positive holding ability of the vertical carriage with the log and dogging assembly running on V-slideways, assures better carriage repeatability through the headrig quad or twin bands. Recovery and accuracy is further enhanced by equipping the band headrig with the Chip-N-Saw closed loop, gas-pressurized, hydraulic high-strain system, which offers most rapid response, excellent damping and constant strain.

The closed loop high-strain system is contained in one block, and pipes, exhaust valves, etc. are rendered unnecessary. To raise the wheel, oil is pumped to the cylinder to take the weight off the wheel, and as gas pressure equalizes to the strain pressure, a check valve closes. There is no return, and the system is fail-safe. Even a sudden power cut-off will not affect the constant strain. Gas volume action cushions the wheel, and the assembly rides. The amount of pre-charge determines the correct amount of damping, and this is adjustable. Any downward movement of the wheel is counteracted by an upward movement of an accumulator piston, and vice-versa, thus the constant strain is maintained.

Table 14.1 is a commercial feasibility study for a

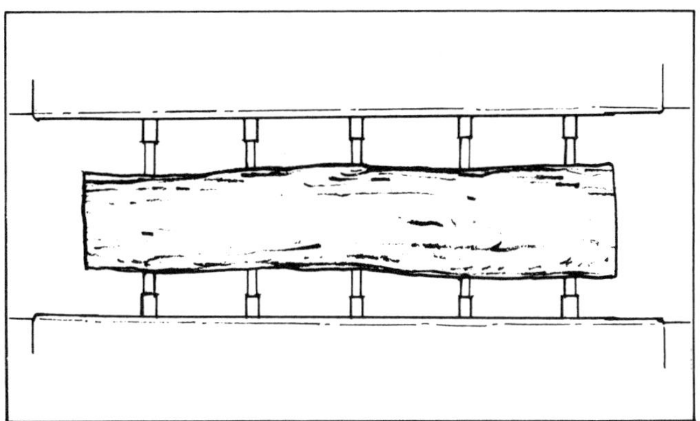

Figure 14.3. Pneumatic cylinders follow the contour of the log to retain firm hold.

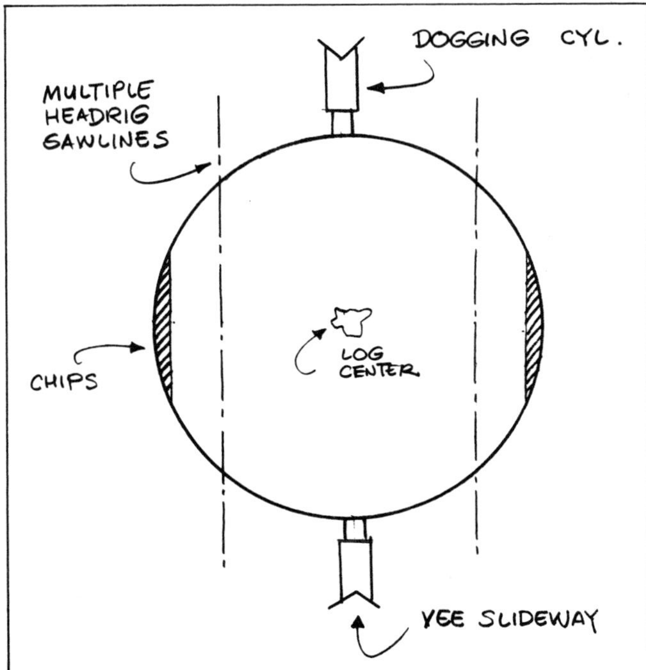

Figure 14.2. Pneumatic cylinders provide top and bottom dogging when the vertical carriage is used.

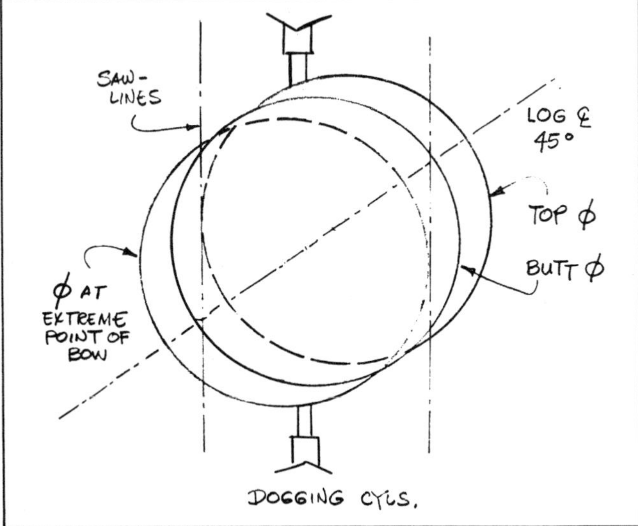

Figure 14.4. The "bow" of the log is reduced when presented on the vertical carriage at a 45 degree angle.

vertical carriage mill in comparison to a conventional carriage mill of North American standards. Mill construction costs, log supply, lumber size recovery, labour and power costs were based on North American standards, but could easily be adapted to Southeast Asian requirements. Probably the major consideration would be that Asian sawmills produce lumber of variable thicknesses, mostly in the 1-inch range, in which case the total board foot production per year would be reduced. However, sales revenue from higher-priced hardwood lumber and lower mill construction costs on buildings and other civil works should compensate on the over-all and keep the percentages on return on investment very similar.

No doubt the innovative principles of the vertical carriage will provoke various questions on the workability of the system. For this reason, the prototype machine has been readied for demonstrations in the Chip-N-Saw plant in Surrey (Vancouver), Canada, where actual processing of Southeast Asian hardwood could be arranged under closely-studied and controlled conditions.

Table 14.1. Commercial feasibility study for a vertical carriage mill in comparison with a conventional carriage mill of North American standards.

A. ASSUMPTIONS

1) Logs range from 10" to 36" diameter and the average tree is 20" diameter and 16 long for a volume of 38.6 cubic feet.

2) The Conventional Carriage with single band headrig can process 500 logs in 8 hours and the Vertical Carriage 50% more.

3) Mill production is in dimension lumber.

4) A further increase of 3% in production by the Vertical Carriage due to more accurate cutting is achievable.

5) The Conventional band headrig lumber recovery factor is 7.5 fbm. per cu. ft.

B. PRODUCTION ESTIMATES (400 MINUTES OF PRODUCTION TIME)

1) <u>Conventional Carriage Production</u>

$$\frac{38.6 \times 500 \times 7.5}{1000} = 145,000 \text{ bd.ft./8-hr. shift}$$

2) <u>Vertical Carriage Production</u>

$145 \times 1.50 \times 1.03 = 224,000$ bd.ft./8-hr. shift

3) <u>Average piece of lumber</u>

1" x 7" x 16' or 19 bd. ft.

4) <u>Average production rate of band headrig</u>

$$\frac{145,000}{400 \times 19} = 19 \text{ pieces per minute}$$

5) <u>Average Production rate of Vertical Carriage</u>

$$\frac{224,000}{400 \times 19} = 29 \text{ pieces per minute}$$

C. CAPITAL COST ESTIMATES

Item	Conventional Mill	Vertical Carriage Mill
Debarker & Cutoff Saw	$95,000	$95,000
Headrig, Carriage, Loaders, etc.	132,000	600,000
Circular Gang	75,000	75,000
Chipper Edger	75,000	75,000
Trimmer complete with Feeder	45,000	45,000
Log & Lumber Handling including sorting	285,000	330,000
Burner	65,000	65,000
Chipping Plant including Blower, System and Bins	200,000	240,000
Chip & Refuse Conveyors	85,000	100,000
Mobile Equipment	300,000	300,000
Service Bay	35,000	35,000
Office & Equipment	35,000	35,000
Roadways	60,000	60,000
Fire Protection	100,000	100,000
Air System including Compressor	30,000	30,000
Electrical Wiring, Lighting Distribution	135,000	195,000
Grinding & Filing Room	30,000	30,000
Mechanical & Operating Spares	25,000	25,000
Building & Foundations	200,000	200,000
	$2,007,000	$2,635,000
Total Including 10% Contingency	$2,207,700	$2,898,500

D. PRODUCTION COST ESTIMATES

1) Log Costs: $75/MBM Of lumber

2) Lumber Price: (rough, green) $120/MBM

3) Chip Price: $30/Cunit

4) Log breakdown at Conventional Mill:

   Lumber: 55%

   Sawdust & Fines: 12%

   Chips: 33%

   Chip Recovery: 0.44 Cunits/MBM

5) Log breakdown at Vertical Carriage:

   Lumber: 56%

   Sawdust & Fines: 9%

   Chips: 35%

   Chip Recovery: 0.45 Cunits/MBM

6) <u>Manpower requirements for headrig mill</u>:

Logdeck	– 1	Trimmer	– 1
Debarker	– 1	Chipper	– 1
Headrig	– 1	Green Chain	–10
Trailer	– 1	Filers	– 2
Edge Chipper	– 1	Millwrights	– 2
Drop Sort	– 1	Electrician	– 1
Circular Gang	– 1	Supervision	– 3

<u>Total Manpower – 27</u>

7) <u>Manpower requirements for Vertical Carriage Mill</u>:

As above, but add the following:

Utility	– 1	Millwright	– 1
Filer	– 1	Green Chain	– 5

<u>Total Manpower – 35</u>

8) Overhead in each mill = $80,000 per year

E.   CASH FLOW PRO-FORMA FOR CONVENTIONAL CARRIAGE MILL
(SINGLE SHIFT, 240 DAYS PER YEAR)

1) <u>Revenue</u>

Lumber:	145 x $120 x 240	= $4,176,000
Chips:	0.44 x 145 x $30 x 240 =	459,360
TOTAL:		$4,635,360

2) <u>Costs</u>

Raw Material: 145 x $75 x 240	=	$2,610,000
Production Labor: 19 at $10,000	=	190,000
Maintenance Labor: 5 at $11,000	=	55,000
Supervision: 3 at $14,000	=	42,000
Parts and Supplies at $1.50/MBM	=	52,200
Power at $0.80/MBM	=	27,840
Overhead	=	80,000
TOTAL:		$3,057,040

3) <u>Gross Profit</u> $1,578,320

4) <u>Capital Expenditure</u> $2,207,700
(This expenditure is for the sawmill
only. The addition of a planer mill
would be reflected in increased lumber
value).

5) <u>Return on Investment</u>

(a) Write-Off period 2 years, 40%

Income Tax Rate   = 45.0%

(b) 20% Capital Cost Allowance,

50% Income Tax Rate   = 30.7%

F. CASH FLOW PRO-FORMA FOR VERTICAL CARRIAGE MILL
       (SINGLE SHIFT, 240 DAYS PER YEAR)

1) <u>Revenue</u>

Lumber:	224 x $120 x 240	= $6,436,800
Chips:	0.45 x 224 x $30 x 240	= <u>724,140</u>
TOTAL:		<u>$7,160,940</u>

2) <u>Costs</u>

Raw Material: 224 x $75 x 240		= $4,023,000
Production Labor: 25 at $10,000		= 250,000
Maintenance Labor: 7 at $11,000		= 77,000
Supervision:3 at $14,000		= 42,000
Parts and Supplies at $1.50 per MBM		= 80,460
Power at $0.80 per MBM		= 42,912
Overhead		= <u>80,000</u>
TOTAL		<u>$4,595,372</u>

3) <u>Gross Profit</u> — <u>$2,565,568</u>

4) <u>Capital Expenditure</u> — <u>$2,898,500</u>

5) <u>Return on Investment</u>

   (a) Write-off period 2 years,
        40% Income Tax Rate = <u>48.6%</u>

   (b) 20% Capital Cost Allowance,
        50% Income Tax Rate = <u>42.0%</u>

6) <u>Return on Additional Investment over Conventional</u>
   <u>Headrig Mill</u>

   (a) Write-off period 2 years,
        40% Income Tax Rate = <u>97.8%</u>

   (b) 20% Capital Cost Allowance,
        50% Income Tax Rate = <u>74.8%</u>

RECAP	CONVENTIONAL CARRIAGE	VERTICAL CARRIAGE
Capital Cost	$2,207,700	$2,898,500
Lumber Production/Shift	145,000 bd.ft. (19 pcs./minute)	224,000 bd.ft. (29 pcs./minute)
Revenue (Lumber & Chip sales)	$4,635,300	$7,160,940
Less Costs Material & Operating	$3,057,040	$4,595,372
Gross Profit	$1,578,320	$2,565,568

(The additional capital investment of $690,800 for a Vertical
Carriage would result in additional gross profit <u>per year</u> of
$987,248).

CHAPTER 15

Use of Edgers in Reducing Waste and Speeding Mill Flow

MARK LAWRENCE
President, Mark 50 Machinery
Portland, Oregon, USA

It is common practice in many areas of the world for lineal ripping of lumber in a sawmill to be accomplished by manually feeding the piece through a single bandsaw. This is usually a slow, inaccurate, and wasteful practice. In most cases, the proper machine for this cutting operation is an edger. The name "edger" in our sawmill vernacular refers to the machine commonly used to cut the third and fourth side to a two-sided board produced on the headrig or primary breakdown unit.

The basic layout

Figure 15.1 shows a typical sawmill. The layout of the mill is simplified and clarified in Figure 15.2. This basic mill process consists of:

1. Headrig—primary breakdown—lumber thickness.
2. Edger—lineal ripping—lumber width.
3. Trimmer—cross grain cut—lumber length.

Figure 15.3 shows cuts made by a headrig, usually a bandmill cutting a log on a carriage. It is not an attempt at best opening face, but is an illustration of a board that requires edging.

Figure 15.4 shows the same board, now lying flat and ready for the second step in manufacturing, the edging process.

Again, the function of an edger is to:

1. Arrive at a predetermined lumber width.
2. Improve grade by eliminating knots, rot, pitch pockets, or other defects, or to separate heart from sapwood.
3. Square up pieces by eliminating the irregular wane side.

The monetary value is thus increased by obtaining the best grade available while at the same time cutting each piece for maximum yield or lumber recovery. The edger saws are used to cut pieces to a desired pre-determined width, such as dimension lumber, or to any width that improves the grade and/or removes irregular sides. We refer to those sides as "wane."

Increasing production

There is much waste potential that develops when cutting tools are operated by people. An edger is no exception. A conscientious, well-trained operator contributes greatly to maximum recovery of rectangular lumber from haphazardly round logs. Conversely, an inexperienced or irresponsible operator makes sawdust or slivers from what could be an attractive, accurate, and, most of all, valuable piece of wood.

The basic edging tool consists of a preparation area or infeed table where the piece is aligned in relationship to the edging saws. This function is most important because it not only predetermines where the saws are to cut, but also gives the operator a chance to change his mind and readjust the board or saws to gain further advantage in grade or maximum lumber yield.

At feed speed averaging 100 feet per minute, the gang saw/edger combination can greatly increase a sawmill's production, while also providing accuracy.

In an exaggerated example of sawing there can be the opportunity to increase production and grade by gang sawing the piece in a gang/edger, which will give you a vertical grain board with spike knots appearing the proper way on the board face (Figure 15.5).

It must be remembered, however, that even those pieces cut on the gang saws might have to be remanufactured on an edger to attain maximum value of the lumber produced.

We have seen how the use of an edger contributes to the milling process by giving more valuable lumber from each log at an increased rate of production.

There are many variations of this secondary milling process to meet the requirements of individual needs, but in most applications, it is the proper second step in converting round logs to marketable lumber.

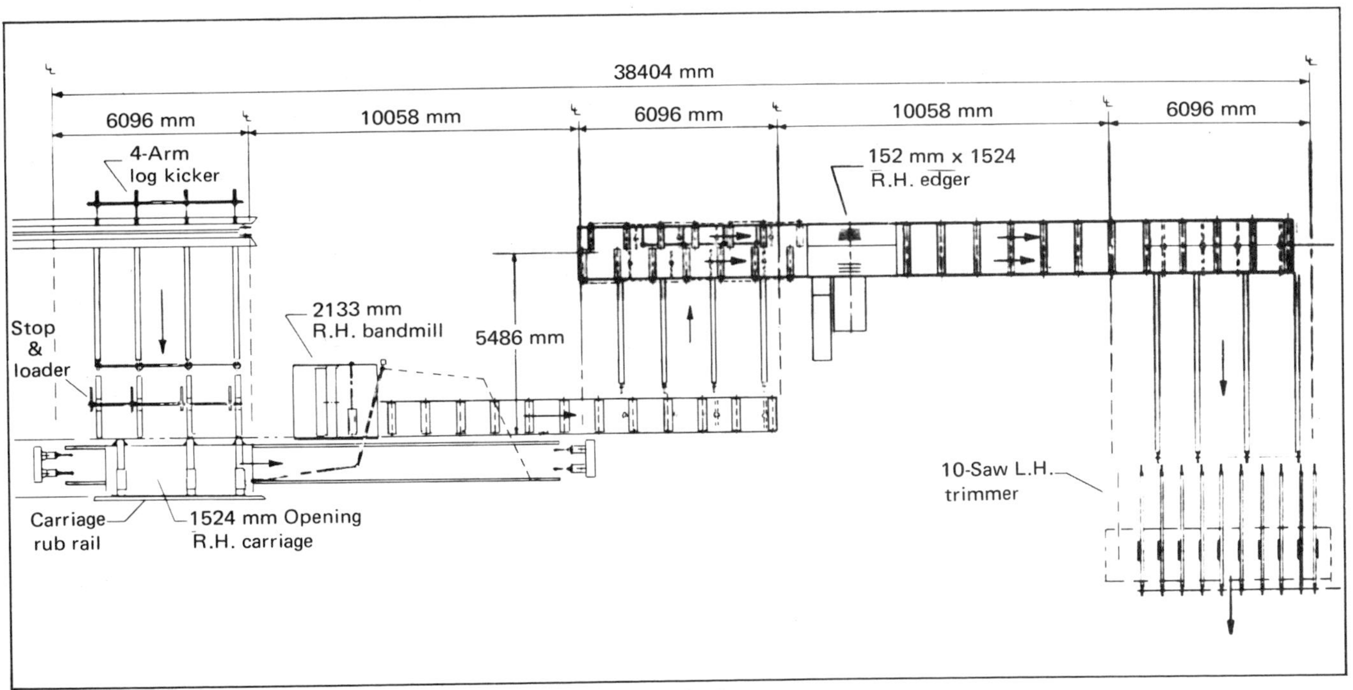

Figure 15.1. A typical mill for cutting hardwood.

Figure 15.2. A basic sawmill layout with headrig, edger, and trimmer.

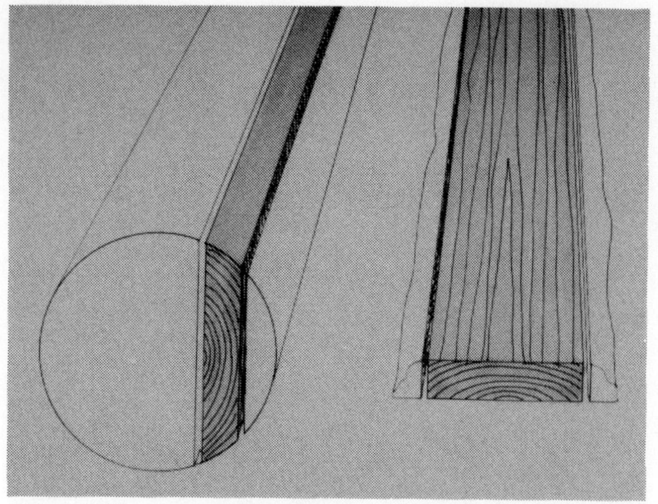

Figure 15.3 Cuts made by the headrig (left), and cuts made by the edger (right).

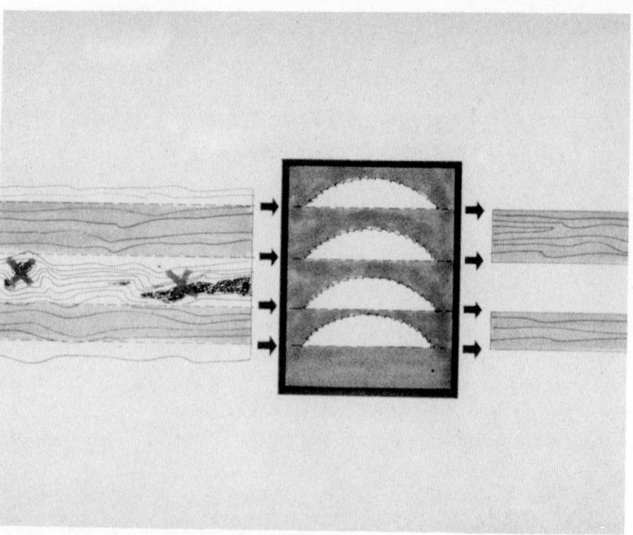

Figure 15.4. Defective wood entering the edger.

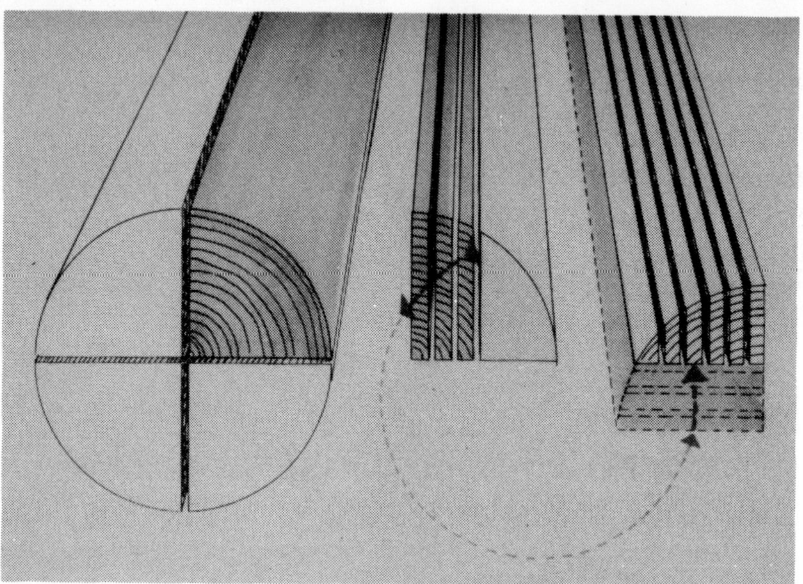

Figure 15.5. A combination gang/edger enables the sawmiller to increase production and grade.

CHAPTER 16

Immediate Profitability from Unsophisticated Machines

RAYMOND C. ISLES
Managing Director, Isles Forge & Engineering Pty. Ltd.
Coffs Harbour, N.S.W., Australia

According to the Oxford English Dictionary, "sophisticated" means "to involve in sophistry, deprive of simplicity, or make artificial." I want to emphasise simplicity. For Southeast Asia I want to present a case, not for innovation or the use of electronics, scanners, pneumatics, hydraulics, and other nameless gadgetry, but for the simplest of mills where top production can be achieved with simple equipment and the reasonable skills of the people in the mill.

I have given this subject a great deal of thought and believe my ideas to be sound. I might just as easily have prepared a chapter on automated sawmilling machinery, in which my company is a leader in Australia, but we are referring to many areas of Southeast Asia, where sophistication is not the answer.

All sawmillers love to talk to one another and happily copy one another's techniques whether right or wrong. They go on holidays and have a wonderful time visiting other mills, and then come home and copy their competitors' ideas, so that a number of inherent faults which frequently occur in many areas of hardwood sawmilling have been reproduced many times. To paraphrase the late Sir George Bernard Shaw—a mistake repeated by many people many times is not any less of a mistake.

Mistakes are made by all of us, but I believe in Southeast Asia, where the strong influence in the past has come from countries whose experience is mainly limited to handling softwoods, many have occurred. Errors, however, can be rectified with some honest appraisal.

In this appraisal we may have a language barrier, not only with the peoples of Southeast Asia, but other countries as well, and this would not be surprising—the same language barrier even exists within the timber industry in my own country. Timber men customarily use different words to name the same thing. Perhaps one word is applied to softwood and one to hardwood. Each timber man has his own favourite expression.

Dating from biblical times when Noah used cubits as a measurement for his ark timbers, measurements and terminology have differed from country to country and from district to district, making communications a continuing problem. We now have increased the confusion with imperial and metric measurements.

We have several names for the same thing; for instance, we call wings roundbacks, we call cants flitches, and we call hearts pith, just to name a few that spring to mind.

Softwood machinery for hardwood operations

It is clear that a great deal of forest machinery in use in Southeast Asia has been purchased on a false premise based on softwood operations instead of tropical hardwoods which are quite different and have unique milling difficulties. This machinery has come, probably in good faith, from softwood-producing countries not familiar with these diffi-culties. This problem applies to mills in Ceylon, Singapore, Malaya, Indonesia, Thailand, Papua-New Guinea, and to a lesser extent, the Philippines.

Improper machinery is very likely to be used in these locales unless a mill is sponsored by a large organisation with its own specialised technical staff, which has information not available to the people in the areas of the mill locations.

Surely it would be practical, then, to have these mills planned so that they are capable of being manned by indigenous people who can understand the simple operation and flow plan, operate it, and maintain it themselves. I surmise the majority of Southeast Asian sawmills operate in rural areas. These areas have some advantages, but because of the rural orientation they are less enticing to highly skilled people who are attracted to capital intensive industries in the urban areas. So Southeast Asian sawmillers must amend their thinking to manage with a staff of less specialised skills.

No doubt, of course, highly skilled people can be imported to the area, but importing personnel would certainly be a costly undertaking, contributing little to the local economy, whereas by simplification, the entire operation can be handled by indigenous peoples.

There are certainly some modern mills in existence in Southeast Asia which are as modern in concept as today, but because of the lack of these specialised skills to operate them, their performance is anything but creditable.

With more complicated machinery as well as that imported from far countries, there is the ever-present problem of spare parts or lack of availability of spare parts. No matter where the original equipment was built, be it Japan, Canada, the United States, Europe, or Australia, getting spare parts takes valuable time, and the more automated the mill the more certain the downtime. Even with Telex, lead time for orders, aeroplanes, and excellent service, the mill could still be inoperative for lengthy periods. Even the use of bandsaws in really remote areas can place the mill owners completely in the hands of their saw filer, and it is well known just how much temperamentalness can generate in this particular section of a mill.

All around the world, I have visited mills where producing 100,000 board feet of timber per day was considered a big operation and a substantial industry, but nowadays, a mill of this size may be only part of an integrated operation with plywood, paper, and other products.

These large complexes will certainly function very effectively with imported technicians and know-how and, in some cases, arrangements are made to train local personnel as skilled artisans for supervisory and management positions. This policy must represent a very substantial and valuable contribution to the host country's economic and industrial consolidation.

The difficulty is finance for these tremendous undertakings, and if we are to consider them as the economic optimum for this region, then there is little hope of an independent, viable forest industry for Southeast Asian countries in the foreseeable future.

The majority of governments in Southeast Asia are now drastically restricting log export and insisting that the finished product be exported instead of the raw material. So you will find small mills will spring up in places which until now have only been remote exotic names. With the vast tropical hardwood forests, timber will, in a fairly short time, provide a wonderful source of revenue to Southeast Asian people. Timber, as we all know is a renewable asset when managed like any other crop, and with the forests carefully husbanded and properly replanted, this frozen forest capital can be transformed into a continuing stream of liquidity to feed many development programmes.

I do not intend to discuss mechanisation of forest operation. Clearly there have been major changes in Southeast Asia in virtually all aspects of forestry practice. Sometimes mechanisation has increased prosperity, but almost always in these industrially developing countries its most obvious impact has been to increase unemployment. Also, the cost of more sophisticated machinery must of necessity over-capitalise the mill. Before I deal with the sawmill itself, I must first deal with the mill owner and management.

Profits

Profit in these days is almost an unknown word. Profits are related to prompt and efficient turnover of stock—all stock. The shorter the period between production and sale, the greater the possibility of profit, except that these greater profits may be quickly eroded by the cost of tied-up capital.

Gone are the days when the sawmiller could simply dump his finished product in the market place or wait for it to be bought, knowing that in time the greater part of the production would be sold. Gone are the days when sawmillers could produce thousands of cubic metres of low-grade lumber and hope to deposit it in the yard of a merchant, while expecting it to be sold at the price he wants.

Reduction and waste utilisation are the motivators of the forest industry, and correctly so. Unfortunately, we in the sawmilling industry have the dubious distinction of being the most wasteful of all the major wood users, but unless we have an integrated complex, this must continue to be so in the future.

Certainly in these days of high interest rates we cannot ignore the proper capital budgeting and the management of cash flows, nor can we ignore the importance of inventory control. We should accept the fact that proper costing systems are essential to our long-term survival, because these systems are necessary for the sawmiller to remain competitive.

Of course, these techniques are fundamental; I did, however, use the word *proper*, and I wonder how many sawmillers, every day, make decisions fundamental to their survival, which are based on inadequate information or crystal ball gazing.

The miller cannot himself control wages, log costs or timber markets but he can control the cost of putting his logs through the mill. Now that we have questioned the mill owner's control of his capital, organisation, and market control, let us return to the mill floor.

The traditional Southeast Asian mill

A breaking down rig, whether circular or bandsaw, followed by a series of small band resaws is not necessarily the best way of sawing timber. I do not know of anywhere except in the Southeast Asian area where this type of mill is installed for hardwood.

Surely then, there is something wrong with this basic concept and I ask "How many Southeast Asian sawmillers have challenged this mill as an efficient production method?" Just make a comparison with other hardwood-producing countries.

An efficient sawmill is not necessarily sophisticated but provides a basic materials handling and production engineering exercise. Once the plan has been thought through carefully, the capital expenditure to ensure the mill's establishment and continued and efficient production can be comparatively modest. I do not believe this careful thought has been put into many of the smaller mills in Southeast Asia, because most of them have apparently just appeared or grown.

One has only to study the inadequate size of many of the machines used. These, in many cases, are not nearly rugged enough to handle the tasks they are asked to perform (Figure 16.1). I pose the question—was it a super salesman who sold these machines, was it the price tag, or were these the only ones available? I believe it was the price tag.

The trend of continuing advanced technology in material handling in the timber industry, particularly in softwood-producing countries, is due basically to the number of logs per hour to be processed and the economic recovery from them as lumber. This, in my view, is not applicable to Southeast Asian tropical hardwoods.

My serious recommendation is to forget sophistication and gadgetry for this area, and revert to the simplest of operations, and I will show you the type of mill I have in mind:

Mill No. 1

1. Log winch, as in Figure 16.2 (hand-operated or electric).
2. Vertical frame saw.
3. No. 1 bench.
4. No. 2 bench.
5. Docking or trimming system.
6. Transfer system between all production machines.
7. Waste system—a green chain or circular sorting table.

A mill of this type, depending on log size and end product to be sawn, can produce up to 60,000 board feet per day in an eight-hour shift.

Mill No. 2

1. Log winch (hand-operated or electric).
2. Vertical frame saw.
3. Circular gang saw or edger (Figure 16.3).
4. Circular resaw.
5. Docking or trimming system.
6. Transfer system between production machines.
7. Waste system—a green chain or circular sorting table.

If the mill owner feels he can handle bandsaws, the circular resaws could be replaced with this type of equipment. The secret of the success of this type of mill is that there are no hydraulics, pneumatics or electronics. All equipment is mechanical and can be handled by service personnel with a small handful of tools. The simplicity of operation is such that an ordinary mechanic or fitter can understand.

The second—and a very important—aspect of the mill layout is the frame saw. It is not designed to just break down a log but to give in the first operation a dimensioned piece of lumber. Other advantages of the frame saw:

- Suitable for all types and sizes of log.
- Gives good recovery.
- Gives straight cuts.
- Gives sized cuts (very important).
- Gives large output.
- Has low saw cost.
- Has low power cost.
- Has low initial cost.
- Is simple to operate.
- Is very rugged and simple in construction giving long service with very little to go wrong.

In Mill No. 1, when typical open or breast benches are used, these machines can be operated by people with very limited education and skills. This equipment, as shown in Figure 16.4, is used extensively throughout the Pacific Islands and Papua-New Guinea. Transfers between machines can be skids or rollers or transfer chains. Again, simplicity is the keyword. Sawdust can be conveyed with water, air, scraper or belt conveyor, while the solid mill waste can be handled by an extractor system. Lastly, the finished product can be conveyed in a green chain, circular sorting table, or board sorter.

Mill No. 2 is basically the same except the circular gang edger will give dimensioned lumber. So from the frame saw through the edger any decision making or problems have been eliminated. All the operator is required to do is to keep the flitch up to the saw. The remainder of the mill is the same as Mill No. 1. Figure 16.5 shows a frame saw handling three logs at the same time. The best opening face, of course, would be the wings on the outside.

This suggested mill design has placed emphasis on manual labour which is in keeping with most government policies, and at the same time, this design eradicates the involved contrivances and devices that are used in the modern sawmill.

To sum up, the Southeast Asian sawmiller should carefully study the layout of his existing mill. Is it traditional in the sense that it is the same as others in the area? Is it the best way of producing lumber or is the mill designed because it has always been done this way? If a new mill is proposed, a decision will have to be made as to whether it is going to be a sophisticated mill with the accent on automation or a simple mill concerned with production techniques and the very best use of labour.

After I had prepared this chapter, I was most interested to read in our Australian Forest Industries Journal of the thoughts of Professor George Marra, Professor of Wood Technology at Washington State University in the United States. The professor was reporting to the third FAO Consultation held recently at New Delhi and I am most gratified that this very learned gentleman agrees with me. He said that what was needed were labour intensive processes and *not* push-button technology. This is exactly what I have been saying.

Let us never forget that in Southeast Asia we have that wonderful commodity—human resources—in abundance. Let us use it to the fullest advantage for the immediate profit and progress of the peoples of Southeast Asia.

Figure 16.1. Large log frame saw with defective log.

Figure 16.2. A log winch, either hand-operated or electric, simplifies the splitting of large logs.

Figure 16.3. The circular gang edger is important in producing dimensional lumber.

Figure 16.4. This open bench can be operated by people with limited education or skills.

Figure 16.5. A frame saw handling three logs at once.

SAWMILL TECHNIQUES FOR SOUTHEAST ASIA

CHAPTER 17

Proper Saw Operations and Effective Saw Doctoring

GERT HARLOFF
General Manager, Vollmer (Singapore) Pte. Ltd.
Jurong Town, Singapore

The working of wooden materials is performed to nearly 100% by chip removing tools. Although the machines on which the woodworking tools are employed are all quite different from a technological point of view—as for instance, circular saws, bandsaws, frame saws, or milling cutters and knives—all such tools have one thing in common: they have cutting edges similar in action and shape for the cutting process.

In spite of the high standard of the cutting tools, it is quite obvious that due to the cutting process the edges get blunt after a certain amount of service. This is not only disadvantageous with regard to the cutting performance and quality but also to the absorbed power and changing of tools, which means loss of time. *It is therefore proved that a successful enterprise owes a lot to the performance of the cutting edges.*

When paying attention to good and trouble-free maintenance for the tools, even with a relatively small investment in tool maintenance machinery, the sawmiller will be guaranteed success, even with rather simple wood-working machinery, when the following points are observed. The cutting edge of a tool should work in such a way that the heel produces—according to the class of tool—either flowing or shearing chips, the size of which depends on the feed per tooth.

Maintenance and care of worn down cutting edges

The wear on the cutting edge is generally greater on the so-called clearance surface. Considering a saw blade as the main tool in the woodworking industry, it is equally important to maintain the lateral clearance of the cutting edges so that the tool cuts freely. When treating a bandsaw or a frame saw the internal stress has to be checked and, if necessary, corrected (see Figure 17.1).

The maintenance of resaws and log bandsaws with a blade width between 80 and 410 mm is considerably more difficult than the maintenance of frame saws and circular saws. In my opinion, a special training of the filer is required because only proper tension and an optimum sharpening of the blade make the bandmill an efficiently working machine. Bandsaws of these dimensions are swaged or stellite-tipped. The tooth shapes of log bandsaws have changed considerably in the course of time, and due to the different grinding principles in Europe and the United States, this development shows essential differences.

Nowadays the important tooth shapes are the so-called skipped tooth and the rounded back tooth which aim:

1. To keep the deepest point of the tooth gullet as far as possible away from the main stress on the tooth face.
2. To keep the radius of the round gullet as large as possible in order to avoid a surcharge by scoring.

3. To avoid marks, scratches, burnt patches as a consequence of a heating-up by the grinding wheel.

The proper selection of the sharpening machine therefore is of decisive importance.

Developments in France proved, in practice, that the following two tooth shapes for these bandsaws are sufficient (see Figure 17.2):

1. The skipped tooth for wood with coarse grain and long fibres.
2. The PCP-tooth with chip deflector and tooth gullet rather far from the face with relatively large rounding in the gullet, with shape of swage and cutting edge geometry varying according to the kind and condition of wood.

When equipping the filing room for bandsaws, attention must be given to whether the bandmill is running in right-hand or left-hand execution. According to the U.S. standard, a so-called right-hand bandsaw corresponds with a left-hand band saw of European standard and vice-versa. We therefore recommend not using these standard definitions when defining the sharpening equipment, but rather determining whether the saw blade shall be fed from the right-hand towards the grinding wheel or from the left-hand towards the grinding wheel when the saw is fitted into the sharpening machine.

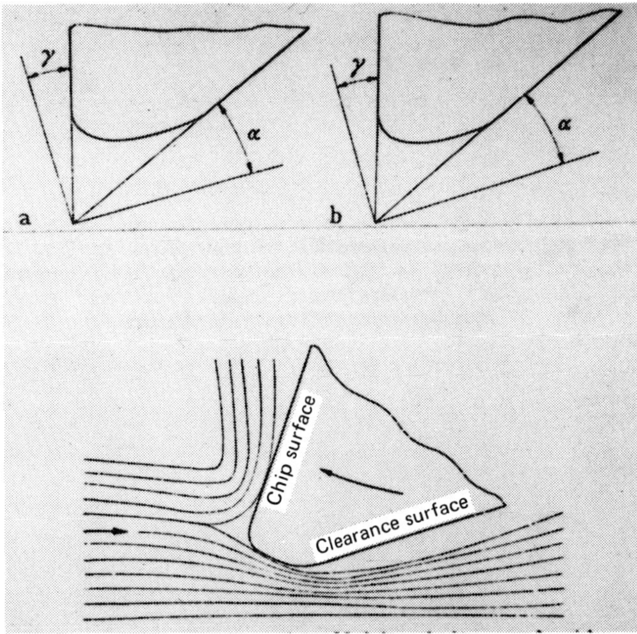

Figure 17.1. Lateral clearance of a saw blade's cutting edge must be maintained so the tool can cut freely.

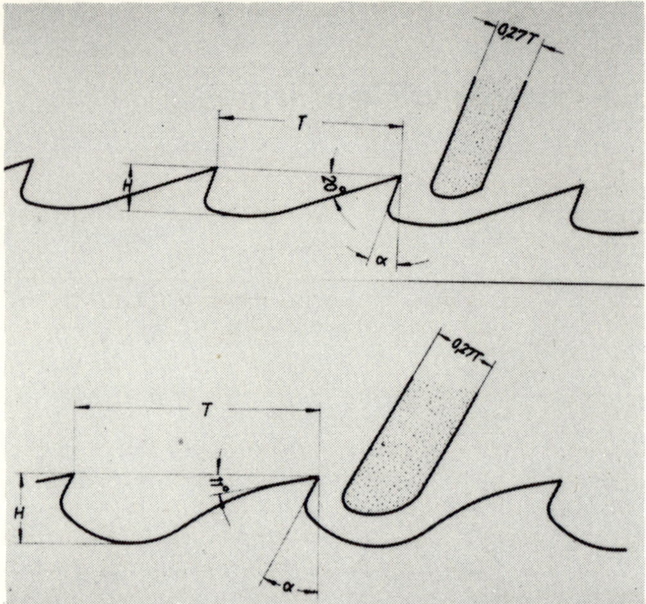

Figure 17.2. These two tooth shapes are sufficient for the operation of these bandsaws.

Machines for the maintenance of saw blades and their special features

Under this heading we want to compare different practices and explain the technical knowledge which we can judge after more than 60 years of experience in this field.

There is an essential difference between the grinding principle as used in Europe and other countires—the so-called tooth form grinding—and the plunge grinding as applied in the United States.

With plunge grinding, the grinding wheel is lowered into the tooth gullet while the saw blade is not moving. The grinding wheel stays there and moves upwards when the feed motion starts. The throat is produced only by the grinding wheel profile. Every tooth shape therefore shows this characteristic rounded profile and has to be re-profiled. The tooth gullet is rather deep, which is disadvantageous for the lateral stability for the tooth.

From a technical point of view, the plunge grinding has the disadvantage that on account of the short moment when the grinding wheel stays in the gullet, this part is heated up to more than 800 degrees centigrade. The surface of the gullet has a hardened zone which increases the risk of cracks. With the tooth form grinding, the movement of the grinding wheel and the feed motion of the saw blade can be varied in such a way that the sequence of the motion corresponds with those tooth shapes which render the maximum cutting results, according to the research of well-known technical institutes.

The tooth form grinding offers the advantages, from a technical point of view, as follows: the movement of the grinding wheel along the tooth face is controlled in such a way that approximately one-third of the grinding movement is used for the finish grinding of the point of the tooth face. Due to the positive inclination of the grinding wheel, scratches and marks are avoided on the face. The movement of the grinding wheel along the tooth gullet and the back is accelerated so that no structural transformation can take place and consequently the formation of heat

cracks is avoided. When grinding the tooth back in the cutting edge zone, the feeding speed is reduced considerably so that, here again as on the tooth face, a finish grinding is attained. When passing the grinding wheel from tooth point to tooth point, a regular wear on the grinding wheel profile can be obtained with a corresponding tooth shape so that the tedious dressing of the grinding wheel is almost unnecessary. The advantages of the tooth form grinding are obvious.

A further difference in the sharpening machines is the construction for the lifting movement of the grinding wheel as shown in Figure 17.3. The left illustration shows a straight-lined, nearly vertical guide of the grinding head with the rotation axis of the grinding wheel during the lifting movement being exactly in the mathematical centre of the saw blade thickness, the so-called "zero-position." By this method, even with wear of the grinding wheel, the tooth gullet and the cutting edge are exactly in right angle to the saw blade plane.

The right illustration shows the system of the lifting lever which does not permit, even with a very precise adjustment of the blade thickness, the tooth gullet or the cutting edge to be straight because it will be inclined to the saw blade plane.

An inclined cutting edge brings these disadvantages:

1. During the cutting process, the saw blade is submitted to asymmetric forces which result in a curved cut as the saw blade deviates.
2. When reswaging, the lateral projection will be uneven.

The inclined tooth gullet has the disadvantage of increasing scratches and grinding marks and favouring the formation of blade cracks.

Tooth pitch and tooth height are the two main measurements of a tooth shape to which the feed and the grinding wheel lift have to be adapted. A sharpening machine on which both adjustments can be made while the machine is running is designed in such a way that the foremost position of the feed pawl and the deepest position of the grinding wheel remain constant even when the pitch and the tooth height are layered. Sharpening machines with these constructional features simplify considerably the work of the filer.

A further important setting is the adjustment of the saw blade thickness which enables the "zero-position" of

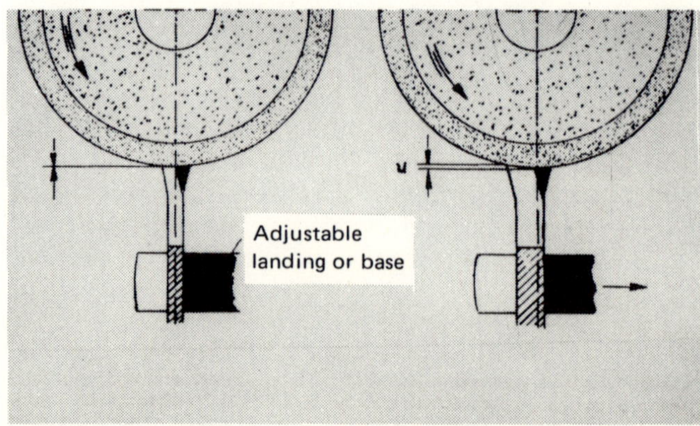

Figure 17.3. Rotation axis of the grinding wheel at (left) "zero-position" and (right) inclined position.

SAWMILL TECHNIQUES FOR SOUTHEAST ASIA

the saw blade to be set exactly under the rotation axis of the grinding wheel (see Figure 17.4). In order to obtain an exact sharpening of the saw blade, a vertically adjustable mounting device is required which should be rigid against distortion and which guarantees that the "zero-position" of the saw blade is maintained exactly in a straight line with the movement of the lifting mechanism. Besides these constructional features, the machine should meet the requirements of enclosed design, stability, and vibration-free construction, which will contribute greatly to a long life of the machine.

A sharpening machine for wide bandsaws with back-feed arrangement and straight-line feed movement is best for bandsaws, as the feed pawl pushes the saw blade in the upper part of the tooth face without changing its position. Only by this method is an impeccable and true tooth shape resharpening possible; whereas feed pawls which push in the tooth gullet will bring no exact tooth face grinding which means uneven cutting width in case of swaged sawblades. Figure 17.5 shows a machine with right-hand feed towards the grinding wheel, but this machine can also be supplied in left-hand execution with feed from the left towards the grinding wheel, so that it is not necessary to turn the wide band saws, whether right-hand or left-hand execution, or to mount these blades in front of the machine without a back-feed arrangement. In place of the back-feed arrangement, a high frequency hardening machine or an equalizing machine could be connected and due to the synchronous drive by the sharpener, the hardening of the tooth cutting edges or the side grinding of stellite-tipped tooth points could be executed at the same time as the sharpening machine is grinding the tooth form. Machines of this type are employed everywhere high requirements are set for the sharpening of bandsaw blades.

Although the principle of a set and a swaged saw is generally known in the sawmilling industry, the advantages and disadvantages of both methods will be discussed in detail. In setting, the tooth points are bent alternatively to the outside, while for swaging, the tooth point is enlarged on both sides (see Figure 17.6). The set saw blade requires at least two teeth to produce the kerf.

The force components resulting from the main cutting pressure and the feed attack the set tooth at a point slightly off the saw centre. In this way, the set tooth is forced outward, and, by the lateral cutting pressure, inward. The swaged tooth is enlarged on both sides and has from the tooth point a clearance angle in the rearward direction and another in the downward direction, that is, in the cutting and in the feeding direction. The force components resulting from the lateral cutting pressure are largely compensated by each other. The main cutting pressure and the feed both attack exactly the blade centre so that the swaged tooth remains in its position. In spite of increased feeding speeds, the cutting quality is better and the deviations of the swaged saw blade are considerably reduced.

When producing the swaged tooth, the eccentric die turns to the tooth face of the tooth to be swaged in the direction of the tooth point while the tooth back is in contact with the anvil (see Figure 17.7). During this swaging operation, the material is displaced towards the sides and towards the point. The swaged tooth must be equalized (see Figure 17.8); that is, the material displaced to the two sides must be compressed from either side to a certain measure so that the tooth has its greatest width in its upper portion

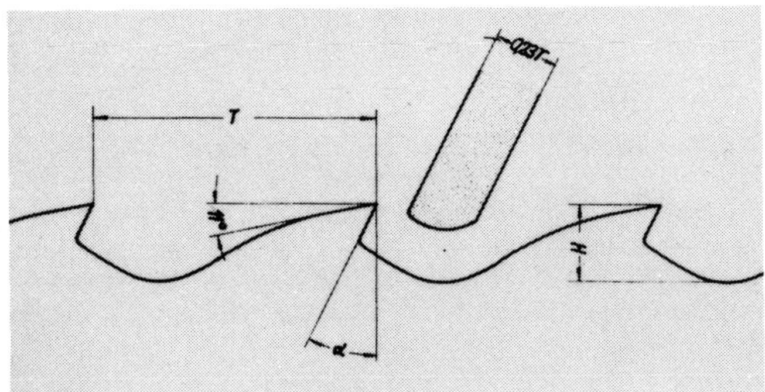

Figure 17.4. Adjustments of the saw blade thickness.

Figure 17.5. Back-feed arrangement with straight-line feed movement.

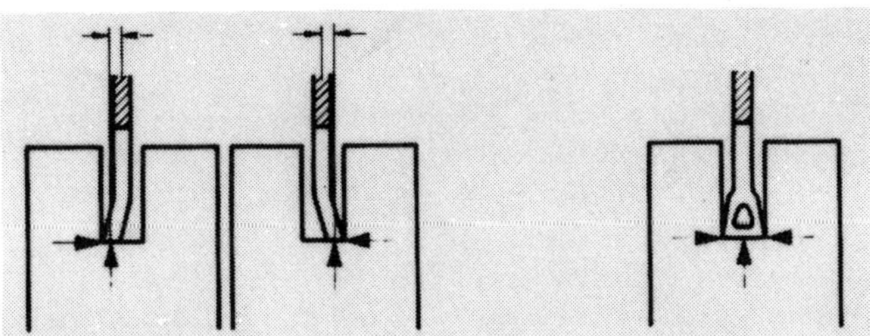

Figure 17.6. In setting, the tooth points are bent alternatively to the outside (left), while for swaging, the tooth point is enlarged on both sides (right).

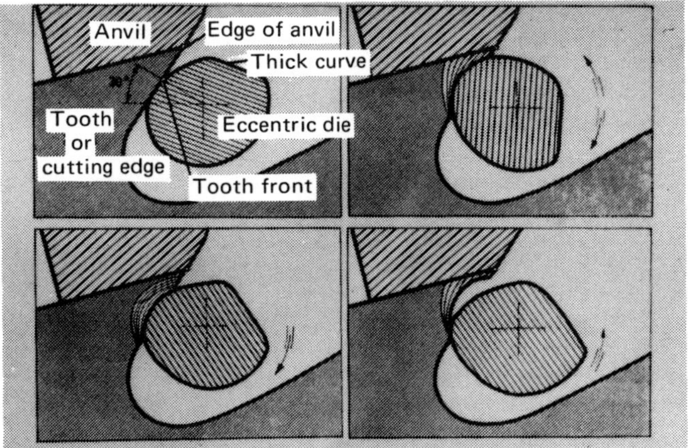

Figure 17.7. In a swaging operation, the eccentric die turns to the tooth face in the direction of the tooth point while the tooth back is in contact with the anvil.

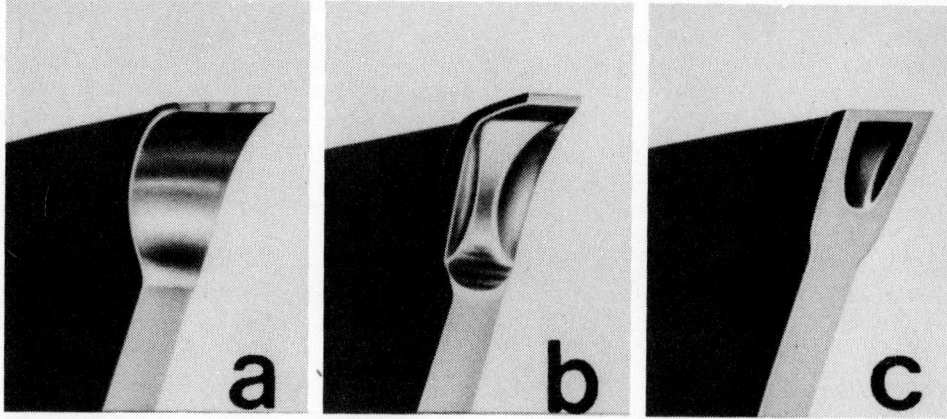

Figure 17.8. The swaged tooth (a) must be equalized (b); then, after grinding, the saw blade is ready for cutting (c).

and is narrowed backwards and downwards. After grinding of tooth face and tooth back, the saw blade is ready for cutting. Owing to the cold shaping of the tooth point, an increase of material strength of 2 to 5 Rc (Rockwell hardness) results, so that the swaged tooth point has a higher strength than the blade body, resulting in a longer service life. For the production of swaged and equalized tooth points, there exist hand-operated devices, swaging machines and fully automatic machines.

There are different sizes and capacities of hand-operated swaging units with various diameters of the eccentric dies. In spite of this, it is usually necessary to swage the same tooth two or even more times in order to obtain the required material deformation. The accuracy of teeth swaged and equalized this way is not always satisfactory, since it depends greatly on the care of the operator.

A further disadvantage of the hand swaging apparatus is the damage of the blade body by the clamping bolts. These bolts have cutting rings at the clamping faces which penetrate into the blade body.

Because of these disadvantages, machines have been developed for automatic swaging and equalizing by pressure of saw teeth. The swaging machine type PTG for blade thickness up to 16 gauge (1.6 mm) executes the swaging and the equalizing in two operations. The swaging machine PMH for blade thickness up to 10 gauge (3.5 mm) executes swaging and equalizing by pressure in one operation. For both machines, the eccentric die and the shaping jaws are separate tools. A combined swaging and shaping tool would have the disadvantage that a certain tooth geometry is given, and the inferior material compaction would result in a shorter service life of the saw blade.

Tensioning, straightening and leveling

Tensioning, straightening and leveling are methods to give a saw blade an internal tension and a geometrical shape required for an unobjectionable working of the saw. This internal tension and the geometrical shape get lost little by little again through the strain of the saw during the cutting operation. For the tensioning and straightening of the saw blade a roll saw stretcher on a tensioning bench is required, and hammering can partially complete the straightening.

Leveling of a saw blade is done, in principle, by hammering both sides of a saw blade using a special anvil fitted in the tensioning bench. Automatic straightening machines

are offered, but up to this date they do not meet the requirements.

Special instructions can be gathered from technical books for the filer but we recommend training the filer at least two to three weeks to get to know the problems and difficulties by his own experience under the supervision of a skilled expert.

After the tensioning of the saw blade, the straight back has to be checked and corrected. The leveling of the saw blade is the last step and should be executed from both sides, and for checking the plane surface of the saw blade, a tension gauge of at least the width of the saw blade should be used.

With regard to the roll saw stretcher, the following points should be observed:

1. Both rollers are driven.
2. The upper part of the machine can be lifted up.
3. The rollers are easy to exchange.
4. The construction is short in order to save space.
5. The adjustment of the guide rollers can easily be done according to scale.
6. The pressure of the rollers is adjustable according to scale.
7. If possible two rolling speeds are available.
8. In case of blade width over 8 inches (200 mm) both rollers are actually movable so that the heavy saw blade does not have to be moved in relation to the rollers.

The anvil plate should be of such a size that the saw blade can not only be checked, but also hammered at the same spot. To move the saw blade to a special anvil and back again for checking is extremely time consuming and troublesome.

Improvement of the service life by hardening

Saw blades for woodworking, such as bandsaws and frame and circular saws are made from steel with a carbon content from 0.6 up to 1.1%, and they can be hardened. The hardening of the tooth points can be divided into two main groups:

1. long-time hardening
2. short-time hardening

SAWMILL TECHNIQUES FOR SOUTHEAST ASIA

The conventional processes such as flame hardening, bath tempering, and resistance hardening belong to the group of long-time hardening. Hardened by high frequency, by plasma current, by ionic beam, and laser beam belong to the short-time hardening group.

The resistance hardening operates with low voltage and high amperage, with the two electrodes fitted to the tooth point and to the saw blade body. As a result of this method, a through-hardened tooth point with a relatively coarse structure is produced. This makes the tooth point very brittle, and there is the possibility that it may break during operations.

With the high frequency hardening for tooth points, the high frequency current is applied only for a short period and only the boundary layer of the cutting edge is heated. Quenching takes place immediately by the coldness of the core material that has not been heated. By this procedure, a hard surface is obtained and it is ensured that the tough core is maintained. The influence of the high frequency occurs intermittently from tooth point to tooth point at no-contact transmission of the electrical energy so that damage to the point is avoided. The hardness obtained by this surface hardening is about 60 to 62 degrees Rockwell.

How does the hardened saw differ from the unhardened saw with regard to performance and service life? A school certificate test at the Fachhochschule fuer Holztechnik (Technical School for Wood Technique) in Rosenheim, Germany, compared spring set log bandsaws and swaged and hardened log bandsaws with regard to service life, cutting performance, power absorption, and cut surface and showed for the swaged and hardened saw blade the following result:

cutting performance	+ 22%
cut surface	+345%
feed	+ 25%

Stellite tipping of band saws

This method is used with best results if hard and abrasive woods are to be cut. Stellite tipped band saws have considerably increased cutting capacity and service life and can be sharpened indefinitely. Apart from tensioning and straightening, swaging is one of the main steps for saw blade preparation, in order to make possible deep embedding of the stellite into the base material. Then the stellite must be applied, either on every tooth or on every second tooth as is often done in order to avoid the removal of the remaining stellite before retipping. This operation is the most delicate one, because it requires skill and care.

The stellite tipping equipment is comprised of a light-pressure oxyacetylene welding torch having a tip with a nozzle of 0.5 to 1 mm (0.03 to 0.04"), an oxygen cylinder with a reducing valve and a pressure gauge, and a similarly equipped acetylene cylinder.

The bandsaw should be passed over guide rollers and through a feeding mechanism. In this way the blade can be fed step by step while the operator is sitting and actuating the foot switch, and neither the welding electrode nor the torch must be laid aside.

The required stellite quantity per tooth is approximately 0.25 to 0.4 grams. Care must also be taken that the stellite is deposited symmetrically. The material stresses produced during the stellite tipping must be eliminated by heating up to a transition zone of about 450 degrees centigrade. For checking, there are special design temperature crayons that change colour according to the action of heat. A skilled operator can carry out this work by his sense of touch only.

Following the stress relieving, the saw tooth points will be given their sharp cutting edge on the automatic sharpeners for tooth form and side grinding. This combination of synchronously working sharpening and equalizing machines guarantees high precision of the cutting edge geometry.

CHAPTER 18

Economic Importance of Proper Tool Maintenance

ERIK W. SUNDSTROEM
Manager of Long Range Technical Planning, Sandvik Steel
Sandviken, Sweden

A 50% reduction in price for last year's models—does that not sound like a profitable bargain? If it concerns tools or maintenance methods, it certainly is not. Even if you can get the old model free of charge, it pays to use the best available tool and maintenance.

Figure 18.1 shows that tool cost is a very small part of the total cost of a company. For Scandinavian softwood sawmills, the average tool cost including maintenance is less than one percent of the budget, compared to 70% to 75% for the logs, including transport to the mill.

For Scandinavian pulp mills, which are usually integrated with sawmills, the cost of chipper knives including maintenance is only 0.2% of the budget compared to 60% to 65% for the pulpwood, including transport to the mill. These figures apply to Scandinavian softwood from second-growth stands. However, the same reasoning would hold for every type of lumber, although with some modification of the numbers.

How the small tool budget is spent has a decisive influence on the company's earnings. Better tools or maintenance equipment are often the best investment you can make to keep down rejects, downgradings and waste.

The problem of fines

In making chips for pulp, the main concern should be to keep the content of fine particles as low as possible. Any wood which becomes fines instead of regular chips means a loss of profit because the fines:

- Give 20% lower yield in cooking when mixed with chips.
- Are the first to be attacked by fungi.
- Are easily blown away from storage piles by wind.
- May cause process difficulties.
- Lower the strength of the paper 0.5 to 1% for each % of fines.

The lower yield of fines means a loss of 0.50 to 0.70 U.S. dollars per ton of pulp for each unnecessary percent of fines, which should be compared to the purchase price of knives, 0.25 U.S. dollars per ton of pulp. This means if a knife gives 0.3 to 0.5% more fines than the best, you should not use it even if you got it free (see Figure 18.2).

Even small increases in fines content mean a big loss if they continue. It is therefore necessary that you check the quality of your pulp chips regularly to verify that the knives are properly maintained. If a 200,000 ton/year mill can lower the fines content 1% by better grinding, this means a 100,000 to 140,000 U.S. dollars profit per year. One other reason for keeping an eye on the fines content is to make sure that the right people get the credit for this good achievement.

Most problems show up early in the chips. Too high a content of fines may indicate:

- Dull or worn knives.
- Too wide an edge angle.
- Worn anvil.
- Worn or loose bearings.
- Deformed wear plates or clamping bolts on chipper disc.
- Obstructions or damaged liner in casing or chip discharge.
- Rotten or very dry wood.

Too high a content of long or oversize chips may indicate:

- Worn anvil.
- Too large a clearance between anvil and knives.
- Too dry wood.
- Inadequate feed-works.
- Too narrow an edge angle.

In-feed problems may indicate:

- Broken knives or knife edges.
- Too narrow an edge angle.

Wear of the knives occurs in two forms (Figure 18.3): abrasive dulling, which is mainly responsible for the fines, and break-outs, which cause infeed problems and rapid consumption of knives. There is a tendency to minimize dulling by using hard knives (H_{RC} 58 = 59) which require care and skill in maintenance, however. The common cause of cracks and break-outs is overheating during grinding which may indicate:

- Too little or misdirected coolant flow.
- Too hard a grinding wheel (grade G, H, I suitable).
- Too heavy a downfeed (max. 0.04 mm per stroke).
- Too slow a table feed (minimum 0.35 meters per second).

Since even small differences in chip quality mean great savings, one has to be very careful when testing different knives or maintenance methods. These are some of the precautions:

- Use only normal grade wood of one species.
- Change knife type each time they are replaced.
- Compare the average value from at least ten chip samples taken at least half an hour apart.

Since the fines content is very dependent on knife dulling, a simpler test may be to compare how far the knives have to be ground to get a fresh edge.

These are some precautions:

- Use only normal grade wood.
- Mix knife types on the disc.
- Grind the knives one by one.
- Compare the total change in width from at least six regrindings, and relate this to the total chip production. The latter routine can also be followed for veneer knives.

Increasing the yield

In sawmills, the yield is also of prime importance, since the value of wood wasted in sawing is usually twenty times the tool cost or twice the machine cost. It is important to remember, that the wood wasted is not only what corresponds to the saw thickness, it is the whole kerf plus the margins needed to compensate for rough surfaces and inaccuracy plus the ends trimmed off because of wane. It is dangerous to believe that the thinnest blade is automatically the most economical. Some figures from Scandinavian mills show that more can be gained by keeping down vibration, inaccuracy and roughness than from using thinner blades (Figure 18.4).

The economical incentive to increase yield varies with the species of wood. Each millimeter of waste when sawing 50x100 mm means 3% less yield or, in Scandinavia, 3 U.S. dollars per cubic meter of sawn goods. For attractive tropical wood, the loss is much more. This should be compared with the saw cost 0.15 to 0.50 U.S. dollars per cubic meter. The easiest way to save is by increasing accuracy, which is partly a matter of personnel training and modern equipment, but neither is sufficient if the tools are not properly maintained.

For bandsaws, the main areas to check are:

- Tensioning
- Wheel crown
- Guides
- Sharpening
- Feedworks

What is wrong can usually be diagnosed from how the band runs.

- The band moves further back on the wheels when sawing: insufficient tension, too small a rake angle, too large a crown, insufficient strain.
- The band moves forward on the wheels when sawing: too large a rake angle.
- The band is consistently too far forward or backward: wrong curve of bands or operator's fault.
- The band runs stable but saws crooked to one side: trough shape from running too far forward, maladjusted guides, or unsymmetrical grinding.
- The band runs stable but is snaking: insufficient tension, unequal tension, badly worn wheels, too small a pitch, too narrow a band, too high guides, or slack in carriage.
- Cracks in the gullets: too much tension, too little crown, too coarse grinding, too small a gullet radius, or too small pitch.
- Lengthwise cracks: too much tension or worn wheels.

One effective method to prevent gullet cracks is to touch the gullet bottom with a rotary carbide burr to remove transverse grinding marks.

For circular saws the main areas to check are:

- Tensioning
- Setting or swaging
- Sharpening
- Feedworks

Diagnosis can follow this plan:

- Rough surfaces: unequal setting or sharpening.
- Crooked sawing to one side: misplaced splitter knife or unsymmetrical sharpening.
- Wavy kerf: insufficient tension, bad feedworks, or overheating.
- Overheating: too small pitch, insufficient swage width or set, improper sharpening, or bad leveling.
- High feed force: too small a rake angle, too small a back clearance angle, too small a pitch, or improper sharpening.
- Fractured teeth: too large a set, too large a rake angle, too large a back clearance angle, or improper swaging.

The total tool cost corresponds to an excess wood consumption of 0.05 to 0.15 millimeter. Many bandmills waste 1 to 2 millimeters through inaccurate sawing, which means 10 to 20 times the tool cost.

Obviously, there is more to be gained from proper tool maintenance than from bargain buying. The filer or grinder has an influence on the economy far greater than is reflected in his tool budget. There is, however, no sense in shortening the regrinding intervals too much. The rate of wear is largest the first hour after grinding, and short intervals thus mean more tool consumption as well as more downtime for changing, which has to be balanced against the better yield or quality. Table 18.1 is a method to find the optimum time between chipper knife regrindings.

DISCUSSION PERIOD

Questions from the floor were answered by a panel consisting of Messrs. Sundstroem and Harloff, and Robert Robertson of McMillan Jardine Malaysia, Ltd., Kuala Lumpur, Malaysia.

QUESTION: Which is best, roller tensioning or heat tensioning?

ANSWER: They both have their places. I think heat tensioning is good when your saw pulls out and you bring your metal right back to where it was before. But you still have to use your rule—you can't ignore it—pull out tight spots.

QUESTION: Mr. Chairman, I'd like to ask three short questions. One, is it possible to side-dress stellite bandsaws on the same machine that you use for sharpening or is it necessary to buy two machines? Secondly, could you indicate the desirable time intervals in relation to sharpening at which stellite teeth should be side-dressed? And, thirdly, could I have some comments on the success or otherwise of refrigerated air guides in the United States?

ANSWER: Let me answer the first question regarding the machinery. According to the system of grinding, it is not possible to side-dress with the same sharpening machine because the movement of the head where the grinding wheel is located differs from the side-dressing machines. On the side-dressing machines, you have two grinding gears to grind, in the same operation, both sides. Actually, we don't deal with stellite-tipped saws on the west coast. I've looked into it quite a bit, however, and I think you can get about ten grinds to a stellite. Then you have to stellite again. As for your question about the air guides, my organization has the water guides, but we haven't got the air. I've never used them. I've never had anything to do with air guides.

QUESTION: My name is Engku Abdul Rahman and I'm from United Furniture of Malaysia. I've heard a number of speakers who spoke yesterday and today and so far there was no mention on the subject of selection of saw blades, whether bandsaw or circular saw blades.

And I would like to make reference to experience locally in Malaysia where we have quite a number of species of timber which are quite abrasive due to high silica content. I'd like to seek the opinion of the panel if you've had such experience and could suggest a selection of blades—either bandsaw or circular saw. Thank you.

ANSWER: On circular saws, I'd recommend carbide. They seem to go quite well in that keruing. For bandsaws, stellite-tipped is the only way to go. For abrasive wood, you should use skip-tooth for bandsaws and we also, as a steel manufacturer, would recommend stellite tipping.

QUESTION: My name is Brian Kilby of Spear and Jackson, Australia, and my question is to Mr. Gert Harloff: In your original recommendation, you recommended, and I again say it was a recommendation, two of your standard profiles and particularly the second one. You show an area with a sharp corner near the base of the gullet. Actually, in manufacturing and maintenance of hardwoods it's shown that any sharp corners have a cracking problem where the saw is used. Were you basing this recommendation on your softwoods of Europe?

HARLOFF: This recommendation comes from France, from a famous institute which is testing the saw blades. So it is not only our recommendation, but it's the recommendation of the CBF Institute in France who are testing, regularly, every kind of wood they can put their hands on. Does that answer your question?

KILBY: No, not completely. I'd just like to add further that we do operate several of your machines and we find them quite successful. However, I would have to be honest in the hardwood area, and certainly honest with the people here, that those two profiles are the ones we avoid because of the cracking problem.

HARLOFF: We have had no experience from cracking at that fairly sharp corner, because it is not at the bottom of the gullet; it is at a place which is not much stressed. And the reason for having that corner is to cause the chips to curl up better so that they will not stick in the gullet. That same type is also used in Scandinavia and Canada for sawing frozen wood or for sawing wood with much pitch in it because that type of wood will otherwise stick in the gullets and not be thrown out.

QUESTION: I'm Manual Bello from the Philippines. In the Philippines, we have several species of wood which have very high silica content and are extremely difficult to machine whether in planing or in sawing. What steps would you recommend to solve this problem?

ANSWER: I would recommend carbide on the circular saws. There are different types of carbide even. For the very abrasive wood you should use the grade of carbide called K-10 which is usually used for particle board; otherwise, the most common carbide is K-20 which is tougher and will stand little pieces of rock and nails better. But for abrasion, grade K-10 is the best. That goes for circular saws.

QUESTION: We cut meranti, keruing, kapur—most of the hardwoods. What peripheral speed would you recommend for the number of species we cut? Do you have a peripheral speed on a band headrig? What would you recommend?

ANSWER: For the soft ones, I would recommend you be up around 9,000 on the rim. But on the hard material I say you should be down about 6,000 on the rim. It all depends on the bite. You can't feed that very fast, so you want your bite larger than the clearance on the side of your saw, so it won't spill by. So you should slow your bandmill down on hardwood, I feel. But on the softer stuff you can feed faster, so you can speed it up. You should try to calculate how much bite you get per tooth and if you go below .2 millimeter per tooth you get excessive wear on the teeth.

QUESTION: We cut five species in one week's time. That would probably mean that we would have to have five different speeds, because your selangan batu is very dense, going up to about 70 or 78 pounds per cubic foot. As against about 44 pounds for the meranti. That's pretty difficult to do—to get a four-speed motor. I've always wished they would come out with a variable speed headrig. That's what they really need.

ANSWER: But you have another possibility, too. You could operate, you could change the pitch on the blade. So you have one blade for each species and change blades whenever you change species.

QUESTION: My name is Huddleston. I was interested to hear Mr. Sundstroem recommend the use of a small rotary file to clean out the gullets of these saw teeth. I once had a saw filer who insisted on rubbing up each tooth after it had been ground with an ordinary slip stone. I objected very strenuously and he in turn objected and we did some costing. I found that with the dressing of the teeth, he was able to get a longer run between grindings than he was without it. Suddenly the accuracy of the saw material was far better with the ground teeth than it was with straight-out ground teeth with the sharpening machine. Another question I'd like to put to you. In regard to these abrasive species, would you like to comment on the hard chroming of the saw faces?

ANSWER: The object of smoothing the gullet was to get the scratches parallel to the steel instead of cross like they are when they are ground. I think you could get some of that effect by filing with a chain-saw file but not as well as you can do with a little rotary file. There are small air-powered rotary machines. You just have to touch a little in the bottom of the gullet. As to chromium-plated versus other saw files, we have tried both at our factory and we have found no advantage with the chromium plating.

We had chromed a set of gang saws where I'm at in Vancouver and it ran well for the first run and then as soon as we ground it once, well, then we were right back with the four-hour change again. But the first run we got eight hours out of it. When you're cutting red cedar out there, it really helps in the wear.

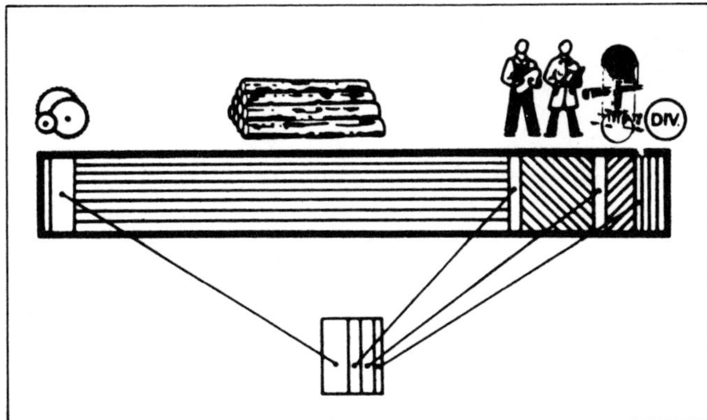

Figure 18.1. Tool cost is a very small part of the total cost of a sawmill operation.

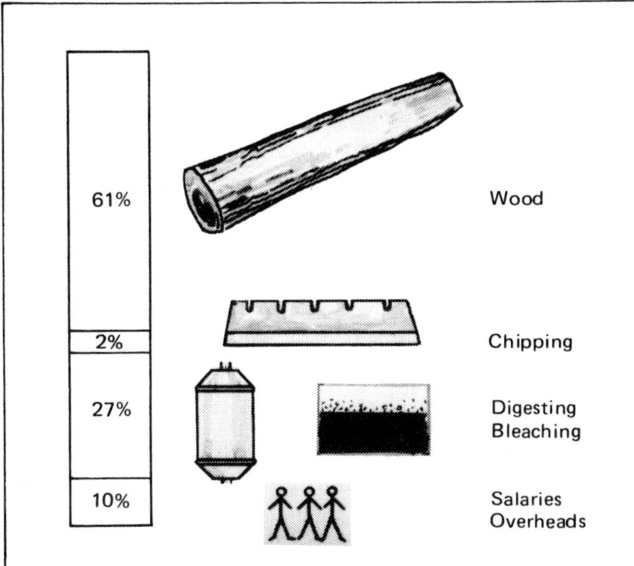

Figure 18.2. Knife economy and sulphate pulp mill costs.

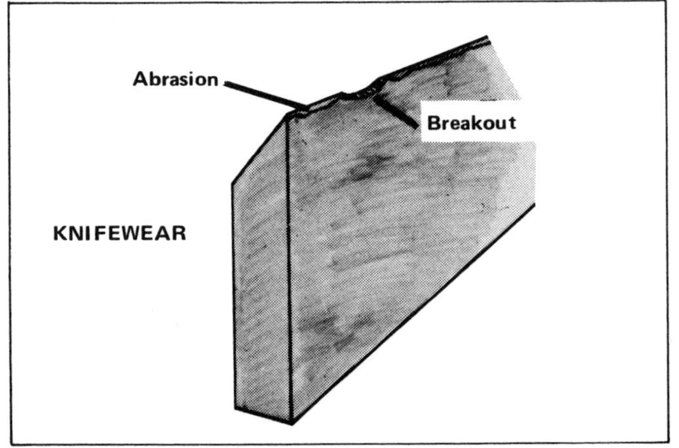

Figure 18.3. Abrasive dulling causes fines, while break-outs cause infeed problems and knife consumption.

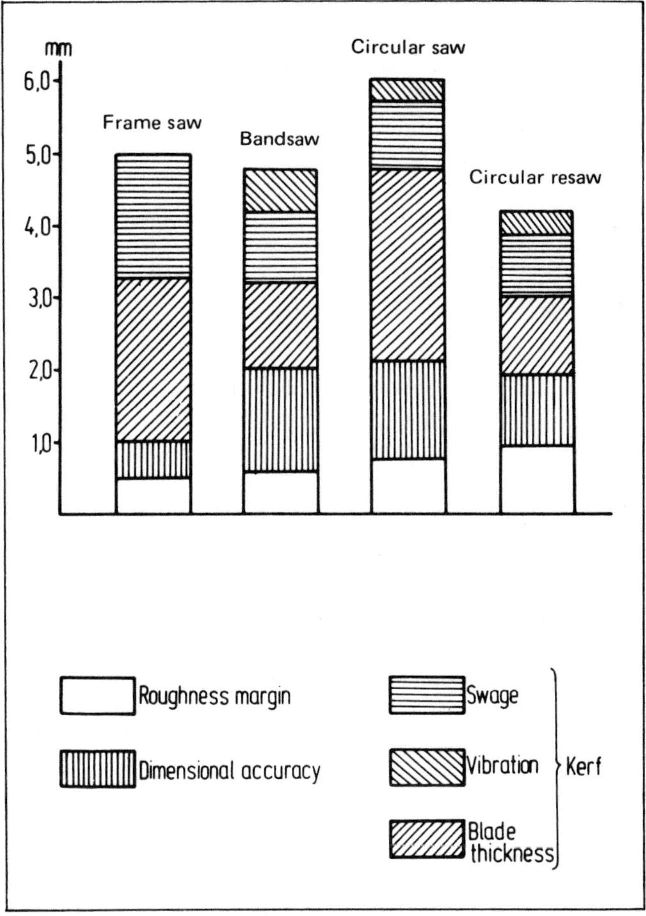

Figure 18.4. Contribution to wood loss from various sources is shown for four different sawing methods.

Table 18.1. A study to find the optimum time between chipper knife regrindings. In this case, there is a flat optimum between 6 and 12 hours.

Interval hours	3	6	12
m^3 chips	1.000	2.000	4.000
Knife consumption US $/ton pulp	0,30	0,18	0,12
Grinding cost US $/ton pulp	0,24	0,14	0,09
Downtime US $/ton pulp	0,48	0,24	0,12
Yield loss (fines) US $/ton pulp	1,40	1,50	1,75
Total cost US $/ton pulp	2,42	2,06	2,08

CHAPTER 19

Mill Alignment, Set-up, and Maintenance

CLAIR A. SMITH
Marketing Manager, Filer & Stowell
Milwaukee, Wisconsin, USA

It is a known fact that no matter how accurate a piece of equipment might be, it can only perform to its claims if the mill is aligned properly. Accuracy is something that every business in the world is trying to achieve. Is there a sawmiller in the world who isn't interested in cutting accurate lumber? No! I'm sure there isn't. The more accurate your product, the better your reputation and the easier it is to sell your product. In short, *there's money in accuracy.*

The lumber industry is a prime example of the effects of accuracy. Boards that are too thick or too thin, cants that are not square, uneven thickness through the length of the cut—all are profit cutters. Not only do you have a bad piece of lumber, but it will probably be handled by your crew through its normal sequence of operations before you find it is not cleaning up at the planer. Along with grade loss, profit loss, and the fact that it takes just as much time to make a bad piece as an accurate piece, we also have a time loss. Accuracy in lumber starts with accuracy in mill alignment. Where should one start to get the accuracy he desires?

Where to start in aligning a mill

The organization uses the sawline as the basis for locating the mill machinery. I am referring to aligning procedures that can be incorporated into an old mill as well as a new mill. It's never too late, no matter how old the mill, to correct your inaccuracies.

The first item I would like to discuss, then, is the equipment used for aligning the mill. We have confined our methods to use materials that are most likely to be found in any sawmill—such as wire, center punches, "C" clamps, etc.

The mill machinery which I will consider aligning for accuracy are the rails, bandmill, shotgun, carriage, headrig chipper, edger, and trimmer. The mill machinery aligned for parallelism are the log deck, log turner and off live rolls. Remember: parallelism = *easy production flow.*

Working with the rails

The "V" rail is the first piece of mill equipment to be aligned. Most people might say they start with the bandmill by setting it up properly and then aligning the rest of the mill to it. However, my organization has found that it is easier to work from the "V" rail, because of its long length, for triangulating back to the bandmill.

Using the mill drawing, locate the rail dimensionally and bolt it down. At this point we are not worried about straightness, but we do want it perfectly level. Shimming may be necessary. Note: the method we use to straighten the rail is best achieved with the use of the track braces. However if your braces do not run full length, some other method would have to be incorporated.

Our method for straightening the rail is as follows

(diagrammed in Figure 19.1), using a wood form approximately 10" long x 6" wide x 3-4" deep or 25 cm x 15 cm x 8-10 cm.

1. Pour slide block, using "V" rail as the model (babbit or lead) and point out approximate dimensions on sheet.
2. When cool, remove excess flashing, then drill hole in center for measuring rod along with a tapped hole for wing nut.
3. Bend measuring rod 90 degrees using a carpenter's square as a guide.
4. Insert into block as shown in Figure 19.1 and hold with wing nut.
5. Weld plates (8" x 3" x ½" or 25 cm x 8 cm x 1 cm) to ends of rail as shown.
6. File notch in welded plate approximately 5" or 13 cm from centerline of V-rail. *Important*—same distance from centerline each end.
7. Stretch piano wire over notches using turn-buckle to tighten.
8. Now mount poured slide block at one end of rail and adjust bent rod to just touch the piano wire. Check other end of rail; it should be exactly the same.
9. Slide guide block along rail using track braces to push or pull rail into line.
10. Relationship of bent rod to piano wire should be the same the whole length.

The flat rail can now be installed. Using the carriage wheel spacing as the width dimension, measure from the centerline of the "V" rail to the centerline of the flat rail and bolt down. Remember, although the flat rail is not as critical to mill performance as the "V" rail, it must be as level as the "V" rail in order to provide a level carriage insuring accuracy of cut.

Don't have a heart attack

The next piece of equipment discussed will be what is the heart of most mills—the bandmill. Let me emphasize at this point that it is important that the bandmill be mounted *on its own foundation and not off of the mill floor.* If the weight of the bandmill should, in time, deflect the mill floor, the result will be a low spot in your rails, which will affect accuracy. The foundation bolts installed by the construction people should position the bandmill in proper relation to the sawline as a first step. The bandmill base should then be leveled in all directions, and then the bolts should be tightened. If possible, weld the keystock stops to the supporting steel structure to lock the bandmill in place.

The following alignment procedure will refer to the lower wheel only (see Figure 19.2).

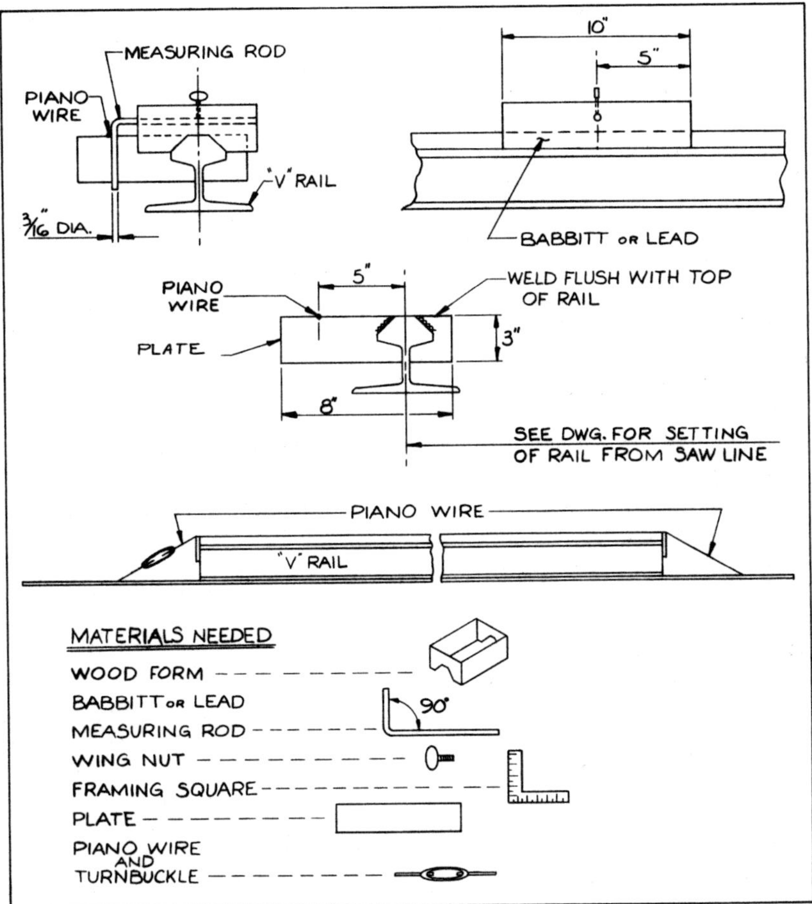

Figure 19.1. Method for aligning the "V" rail.

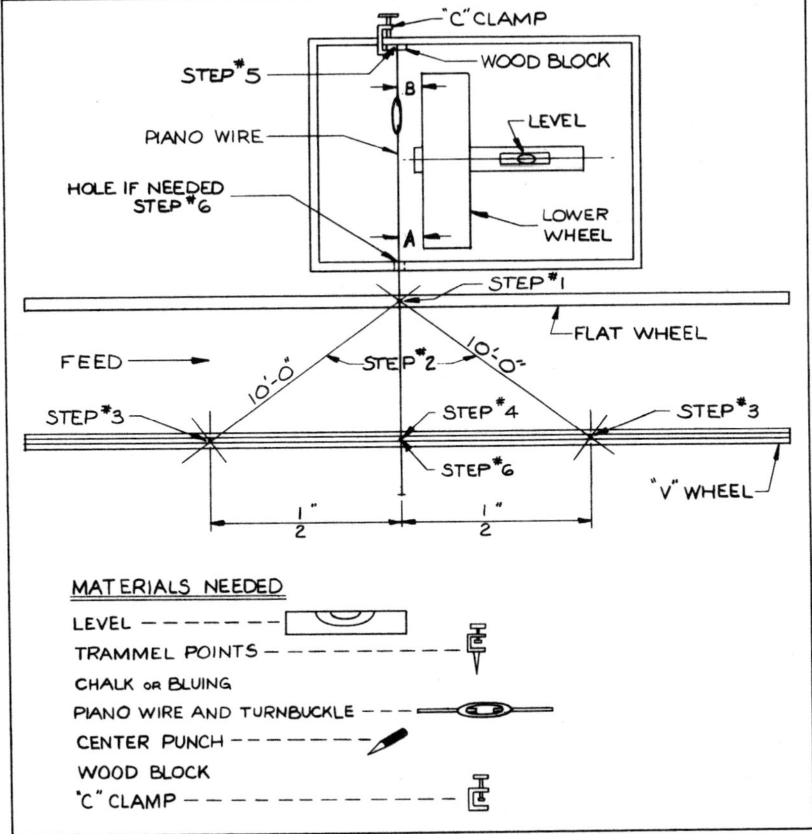

Figure 19.2. This alignment procedure refers to the lower wheel of the bandsaw.

1. Center punch a mark on centerline of flat rail approximately 6" or 15 centimeters away from wheel edge (feed side).

2. Using trammel points (or if nothing else is available, a straight stick with a nail in each end) use the center punch mark as a pivot, and scribe a line on the top of the "V" rail (layout bluing would be a big help here) in both directions.

3. Punch a mark on the "V" rail where scribed line crosses center line of "V" rail.

4. Measure distance between two punch marks on "V" rail. Set trammel points to ½ the measured distance. Now scribe from each punch mark on "V" rail toward center of "V" rail. Punch a mark where scribe marks cross the centerline of the rail.

5. Attach piano wire to back of bandmill base with block of wood and "C" clamp.

6. Stretch piano wire over center punch mark on "V" rail and tie down. (Note: depending on bandmill base location, in relation to top of rails, a hole may have to be drilled in base to allow wire to pass through.)

7. Move wood block (which is "C" clamped on back of base) until wire is on centerline of punch mark on flat rail. Wire is now at a perfect 90 degree angle to "V" rail.

8. Measure over to lower wheel and adjust wheel until measurement is the same both front (A) and rear (B).

The lower wheel is now adjusted horizontally. To adjust the wheel vertically, mount a good machinist level on the lower wheel shaft and adjust until level. The next step is to align the upper wheel to the lower wheel (Figure 19.3). The upper wheel will first be checked horizontally.

1. Attach two pieces of wood to outer surface of upper wheel with a "C" clamp about 12" or 30 cm above center.

2. Tie off two plumb bobs, one each side, about 6" or 15 cm out from edge. Let them extend below bottom wheel.

3. Measure at "A" top wheel both sides to read the same. Then measure at "B"—this should read the same as "A"; if not, move upper wheel until measurement is the same. Now make vertical check.

4. Rotate one stick and "C" clamp to top of wheel. Plumb bob line should just clear shafts and extend below bottom wheel.

5. Check dimension at C-D-E-F—all should read the same. If not, adjust top wheel accordingly. At this time, remove all attached equipment.

6. Take one of the plumb bobs and tie it around upper wheel as shown. Let it extend down front of both wheels.

7. Line up top wheel to bottom wheel face. At this point, wheel alignments should be perfect.

8. Put saw on, then properly strain.

9. Align upper saw guide to saw and tighten.

10. Adjust top wheel straight back to achieve proper pressure angle (usually about 3/8" or 1 cm).

11. Adjust bottom saw guides at this time. When bandmill is started up, the wheels may be adjusted to "train" the saw and adjusted for lead.

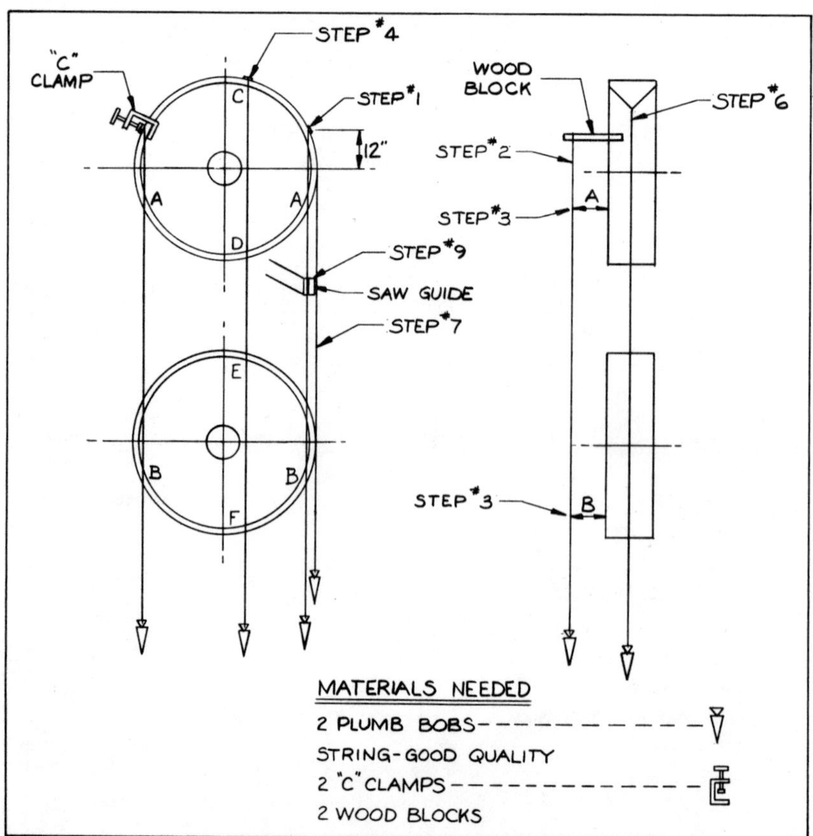

Figure 19.3. The bandsaw's upper wheel should be aligned to its lower wheel, as shown.

Plumb bob tips

At this point I would like to mention a word about using the plumb bobs. If at all possible, don't use a bolt and a piece of twine; but try to get two good plumb bobs (mercury-filled are the best). It helps to let the plumb bobs hang in a container of oil while being used; if someone bumps one it doesn't take a long time for it to come to rest. Also, if the string is fuzzy at your measurement point, rub it with a bar of soap. This will give you a smooth string.

Checking out the feed alignment

In lining up the shotgun we again use the "V" rail as a guide (see Figure 19.4).

1. Bolt separate sections together as you bolt them down to anchor bolts. Leave anchor bolts loose.
2. Install piston and retract.
3. Using dimensions shown on drawings, adjust shotgun for vertical height. Check at piston rod face and at rear section of last piece. Shim if needed.
4. Again using dimensions shown on drawings, measure from centerline of "V" rail to centerline of shotgun, at piston rod and last section.
5. Attach piston rod to carriage swing bracket.
6. Now tighten anchor bolts.
7. Attach necessary plumbing and operate gun slowly through several cycles.

Checking out the carriage

When aligning the carriage, it is important that all "V" wheels contact both sides of "V" rail (see Figure 19.5). You should not be able to get any feeler gauges between wheel and rail. If you can:

1. Loosen axle mounting bolts and adjusting screws on trucks that need adjusting.

2. Run carriage back and forth a few times.
3. The wheel should now have centered itself.
4. Tighten all bolts.

The next item that should be checked are the knees. Run the knees back and forth a few times and activate tapers.

Important: Make sure there is no water in the taper cylinders from the air system. This will affect the accuracy of alignment.

1. Run knees to forward-most position.
2. Locate first knee opposite saw teeth.
3. Measure from designated tooth on saw to face of knee using inside calipers.
4. Check all knees with same saw tooth.
5. If variation is found, adjust knees by adding or removing shims at rear of knee. Carriage alignment should now be perfect.

Next the chipper

The headrig chipper is one of the easier pieces of machinery to align (see Figure 19.6). Begin by loading a log on the carriage and taking a saw cut off the log.

1. Activate chipper cutter head cylinders so that cutting head is in the forward or zero position.
2. Bring chipper cutting face knife up to and touching the log, by moving chipper base.
3. Adjust lead by rotating designated facing knife from cutting position, 180 degrees.
4. Using a feeler gauge, adjust for a 1/64 lead.
5. Now level machine in all directions and bolt down.

From this position, any adjustments can be made to final alignment to sawline by using the limit switch setworks.

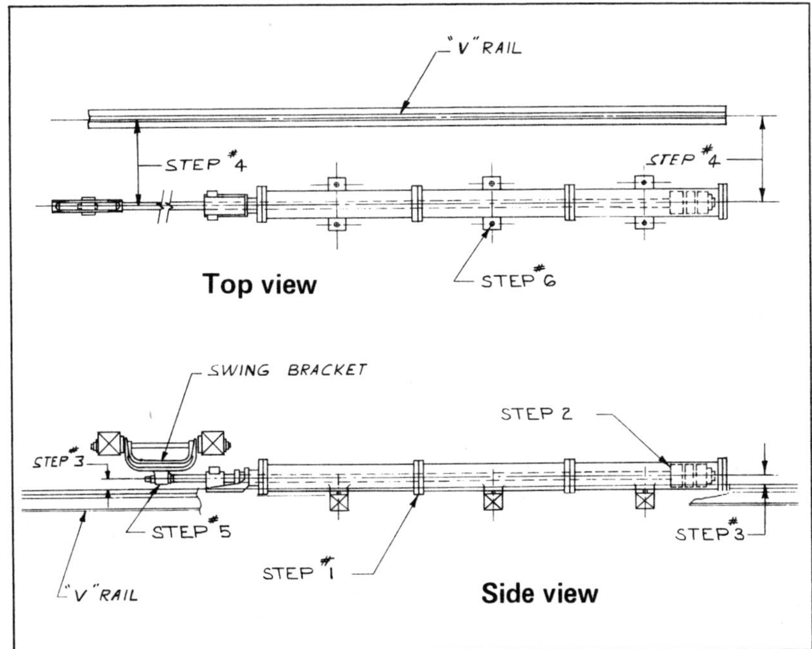

Figure 19.4. Procedure for aligning the shotgun.

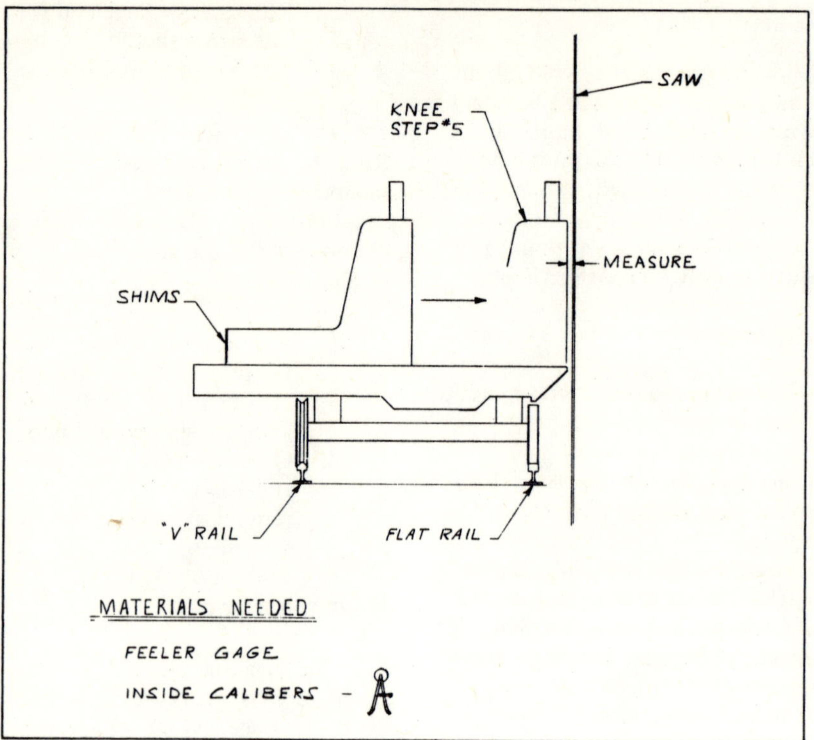

Figure 19.5. Method for aligning the carriage.

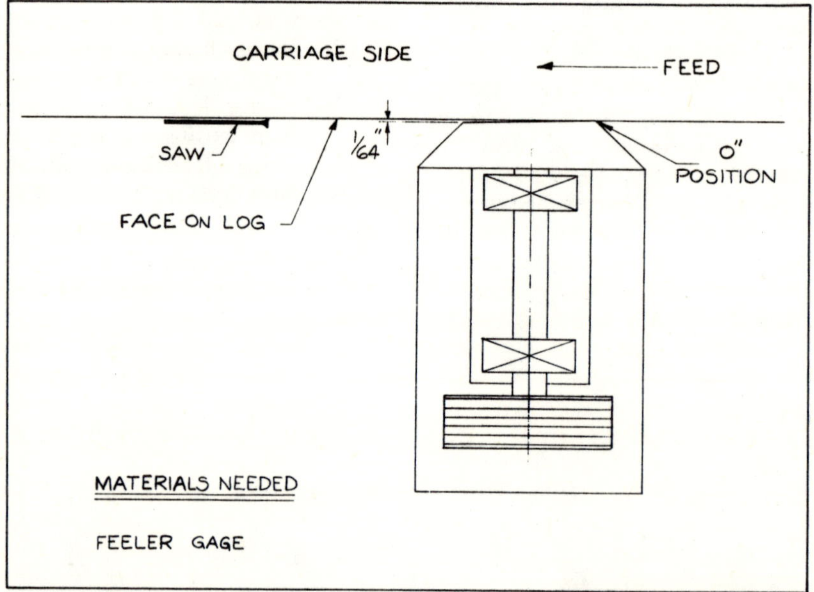

Figure 19.6. This is the alignment procedure for the headrig chipper.

Other equipment

The alignment of the edger and trimmer are basically the same, in that the infeed rolls or infeed chain, as the case may be, must be perpendicular to the line of the saws. Infeed equipment to any saw must come in at a perfect 90 degrees. Anything other than this will cause binding and miscut lumber. The log loading deck, the log turner, and live rolls should all be checked for parallelism to their related machinery.

Now that we've aligned all the equipment, the question that next comes up is how often should the alignment of the mill be checked? I would recommend the following:

1. Every six months. Over a weekend or during a holiday shutdown, check out one or two items until the alignment of all machines in the mill has been all checked.

2. After every accident. If a log should break free of the carriage and jam the chipper or bandmill, be sure to check for misaligned or bent equipment and correct it immediately.

3. If at any time it becomes obvious that the lumber is not sawn accurately or has lost its quality features, and it has been determined that the error is not a saw blade or setworks problem, then recheck alignment of the mill machinery.

CHAPTER 20

Proper Dry Kiln Operation

ALI ZAFAR
Managing Director, Hildebrand Singapore Pte. Ltd.
Jurong Town, Singapore

Southeast Asian exports of sawn timber have increased steadily in the past two decades with the last ten years seeing a major and rapid growth in this field. With greater and wider acceptance of the Malayan grading rules in countries of importation, setting up of larger and better sawmilling plants in the region, and awareness of the properties of some previously lesser known species, this progressive export trend has all the potential for a steeper rise. It must be noted that the end use to which the timber for export is applied determines to a great extent the species, sizes, and properties of the sawn wood. With the increasing price and growing difficulties in procurement of good quality sawn wood, there is a tendency to utilise imported timbers as raw material for more expensive construction only.

Higher ocean freight, rising wages, and problems of waste disposal on the end of the overseas buyers are other factors which demand importation of semi-finished and more standardised raw materials. Timber being not a man-made material, it is almost impossible to ensure that the complete production from a sawmill be entirely free from defects for the purposes of each buyer. Great efforts, however, are spent by sawmillers and timber exporters to ensure a minimum of rejects and claims. Kiln-drying of sawn timber is one operation that goes a long way in complementing a move for standardising the quality of a sawmill production and obtaining the trust of the buyer for a long-lasting trade relationship.

Kiln-drying is a prerequisite for most timber processing operations beyond sawmilling which are based on solid wood—for instance, timber mouldings, floor parquet, and quality furniture manufacture. Contrary to such industries, however, sawmilling is an industry where a plant can thrive without necessarily having a kiln-drying facility of its own.

Kiln drying and the buyers

There is a ready and established demand for graded and ungraded air-seasoned timber, because many buyers kiln-dry the timber at their end. This probably is the main reason which is keeping many sawmillers from going into kiln-drying. No doubt, not all timber converted to sawn boards needs to be kiln-dried prior to use. However, the percentage is very much higher than the approximately 10% of all sawn timber exports which emanate from Southeast Asia in kiln-dried form. Of about 42,000 cubic metres of total sawn wood exports from Malaysia in February 1975, for example, only 5,000 cubic metres were kiln-dried. Similarly, Singapore exported a mere 51,000 cubic metres of kiln-dried lumber against a total of 820,000 cubic metres of total lumber exports in 1974. Indonesia exports little kiln-dried timber and figures for the Philippines, which are not readily available, are not too different.

The reason why buyers abroad do not so often specify kiln-dried timber in their requirements is, in fact, due to the relative difficulties they have in obtaining sufficient quantities of kiln-dried lumber. A trend overseas to buy more of kiln-dried lumber rather than rough-sawn, air-seasoned timber has been growing over the past few years, and it is expected to gain further momentum with the recent escalation in operational costs of kilns in Europe, the United States and Japan. At the same time, sawmillers in Southeast Asia are now slowly realising that they can obtain premium prices for their lumber if it is kiln-dried and if they can build up buyers' confidence. The chances are, if you are not kiln-drying, you are probably selling much less than what you could, obtaining low prices for your valuable material, and achieving a slower growth rate for your sawmill.

A simple operation

Kiln-drying is one of the simplest of all timber processing operations. A process of water removal alone is involved. In fact, drying of timber is a natural process, which, under normal conditions, one would need special efforts to stop. All kiln-drying does is to accelerate the rate of drying with artificial aerothermal applications and control to avoid defects which normally result in air seasoning. Information on proper operation is usually readily available from established manufacturers of equipment who maintain well-equipped laboratories for the purposes of developing suitable drying schedules for various species of timber and who have results of actual dryings available from almost all parts of the world.

It would be interesting to examine the various details that one will be involved with in and before the setting up of a kiln plant. I will try to describe these in maximum detail without making the text unduly complicated.

Drying requirements

When one starts looking up suitable kiln equipment, some necessary data is to be compiled before approaching a manufacturer of dry kilns. Basically, there are three factors which are important:

- The total load or, in other words, the quantity of timber in cubic metres you wish to dry per month.
- The load size per kiln or the net holding capacity of the chamber which means, in other words, the quantities of the timber which can be successively placed inside the chamber in the form of a proper stack for drying.
- The total time it will take to dry the load of timber placed inside the chamber.

In most cases, the amount of timber you wish to dry per month would be easy to calculate. The size of the chamber and the drying time are interlinked, and in order to understand this problem, we shall now make some assumptions initially.

Holding capacity

Let us assume that the drying time for a timber in question is five days. This means that once the timber load is fed inside the chamber and the door closed, the load can be taken out after completion of drying, i.e., the chamber would be occupied for five days. With 30 days to a month for example, this would mean the chamber can be loaded six times per month. Now if for example, 1,200 cubic metres of sawn timber is to be dried every month and if six charges can be dried a month we would need each charge to be 200 cubic metres in order to get all the timber dried within a month. A kiln of 200 cubic metres holding capacity is usually too large and, therefore, several smaller chambers would be installed which have a cumulative holding capacity of 200 cubic metres.

Importance of drying times

Thus, it can be seen that knowledge of the speed at which a certain timber can be dried is essential in order to decide the size of kiln equipment to be purchased. This brings us to the question of how one can know the drying time. Normally, manufacturers of kiln equipment will be able to provide such information. It is highly recommended, however, to make the estimates on your own based on the specific requirements in your own case. The drying time of timber depends on several factors, most important of which are:

- The *species* of the timber to be dried.
- The *thickness* of the boards which are intended to be dried.
- The *initial* moisture content of the timber, that is, the amount of water in the timber planks before being dried.
- The *final* moisture content to which it is desired to dry the timber. This is related to the purpose for which the timber is to be used and climatic conditions of the *location* where the fabricated goods will be placed.
- The *use* to which the dried timber will be finally applied, that is, whether it will be used for building construction, furniture, sports goods, or musical instruments. These end uses determine the tolerances in *quality* of the dried timber which are acceptable.

With the above data in hand, it is possible to calculate the drying time and, thereby, the size of the kiln plant more accurately.

Timber structure and thickness

Mainly, the speed of drying varies from timber to timber due to the differences in the cellular structure of the wood. Timbers which are dense in structure and have thinner capillaries require a longer time to give out their water. These timbers are generally heavier in weight also; on the other hand, timbers with a simpler cellular structure, which allow the water to travel to the surface more easily, give up their water much faster and, therefore, require a shorter drying period. Of course, even for the same timber two boards of different thickness will require different drying times within the same limits. This is due to the fact that timber dries in a kiln by evaporation largely from the top and bottom surfaces of the plank. Since water in a plank has to travel to the surface in order to be carried away by the air flow, the thicker the board of timber the longer it takes the water to appear on the surface; besides, movement of the water from the inner portions to the surface results from a gradient set-up between the core of the timber and the surface due to the difference in temperature and air movement. This gradient is also affected by the thickness of the timber.

Initial moisture content

If a kiln is set up in the forest and the boards to be dried come from freshly sawn logs, the initial moisture content of the boards would normally represent the usual moisture content of the tree trunk itself. It is interesting that the amount of water held within a tree trunk varies very widely from one tree to another. For example, while the moisture content for a certain species might be 60%, in another case it may be as high as 200%.

Moisture content definition

A proper definition of moisture content at this juncture appears necessary. In timber technology, moisture content given in percentage represents the amount of water in weight as compared to the amount of bone-dry solid wood component in a piece of timber; that is to say, if we have a piece of timber which weighs 2 kilograms when freshly cut from a tree, and on drying to an extent when there is no water left in this piece, the weight would be 1 kilogram. The moisture content of the original specimen will be expressed as 100%.

Final moisture content and use

Timber is, however, never dried to the extent that no water whatsoever is left in it, that is, to 0% moisture content. There is also no standard final moisture content to which all timber requiring kiln-drying has to be dried. As was pointed out earlier, the purpose for which the dried timber will be used and the location where the fabricated item will be placed determine to what moisture content the timber must be dried. Timber is generally dried to ensure that it will not undergo any changes in form, colour or size once it has been converted to its final shape. Various tolerances, however, are permissible depending on the final product. For instance, a slight shrinkage on checking or warping may not be critical in timber used for building construction. However, timber that will be shaped into expensive furniture, wall paneling, floor tiles, or musical instruments ought to be dried to such an extent that no changes would occur for several years to come after the goods have been fabricated. Even in the simpler case of a window or a drawer, shrinkage during drier seasons and expansion of windows during more humid seasons can result in windows that cannot be closed or drawers that are stuck.

Drying time and use

Also, the expensiveness of your timber will influence the drying time, because you cannot use harsher drying schedules that may result in rejects. Harsher schedules can also not be used for timbers where the final dried timber is required to be absolutely free of any defects.

Final moisture content and location

The location at which the timber will be used affects the final moisture content for drying due to the fact that the climatic condition has a great influence on the moisture content of the timber. If a timber is not dried down to the right moisture content desirable in a certain climatic condition, it would either continue to give up some moisture content or take up some water from the atmosphere. This poses the question of how one would know the moisture content of the wood which would be in equilibrium with the climatic conditions of the ambient atmosphere. In fact, the moisture content of wood, at which such a condition will exist is called "the wood equilibrium moisture content."

Equilibrium moisture content

The temperature and relative air humidity of the location in question is directly related to the wood equilibrium moisture content. A relationship has been established between the wood equilibrium moisture content and the two factors determining climatic conditions—temperature and relative humidity—and is available in the form of charts and diagrams. With the help of such graphic diagrams, one can immediately look up the equilibrium moisture content for certain specific climatic conditions. In places where climatic conditions differ considerably over the period of the year, the equilibrium moisture content also varies accordingly. For example, while in New York, the equilibrium moisture content varies between 7% to 13% from January to July, in a place like Jakarta the equilibrium moisture content would be about 15% almost the year round. Again, even in climatic conditions prevailing in Southeast Asia, the wood equilibrium moisture content would be very much lower in places where climate is artificially controlled, as for instance in air-conditioned buildings and rooms. Knowing the equilibrium moisture content of the location where the finished timber produce will be placed, it would be necessary to kiln-dry the timber down to that moisture content.

Initial moisture content

As for the initial moisture content of the boards to be dried, it should also be noted that the moisture content would not always represent the original moisture content of the tree trunk. While there is usually a big lapse of time after the tree is felled and brought to a sawmill there is, however, very little loss of water from this log due to the massive volume and shape of the log as well as the fact that the bark still surrounds the tree trunk. As soon as the log is sawn into boards the drying process starts inevitably due to the large surfaces exposed to the atmosphere. Very often sawn timber is stored or is in transit for long durations before it is charged into kilns for drying. In many instances, sawn timber is properly stacked and allowed to dry naturally with the help of the ambient temperature and air flow. Thus the timber planks lose considerable moisture content during this process and are dried down to a certain extent (to about 50% to 30%) before being kiln-dried.

Calculating drying times

After gathering the data necessary for the determination of the drying times, one has then to set about calculating this value for the particular timber in question. Although empirical formulas exist for the calculation of drying times, a more practical approach is to refer to actual drying results which exist for most of the timbers of interest. Such observations are usually recorded graphically as a ratio of wood moisture content to time. This relationship is almost parabolic and one can read the time required to dry a certain species of timber within the desired moisture content limits. This graph is based normally on the result of drying according to a certain drying schedule. This drying schedule, of course, is designed for the type of drying required which takes into account the purpose for which the timber is intended and the permissible degrade in or damage to the dried timber. A kiln-drying schedule is in fact a table setting out the temperatures and the relative humidities of the air inside the chamber appropriate for drying a given species of timber to a certain moisture content.

Drying schedules

In order to avoid the necessity to change the conditions continuously while the timber is drying and changing its moisture content, it is customary to draw up a drying schedule in definite phases or steps, with each step representing a certain fall in moisture content. Various drying schedules exist for the same species and thickness of timber, depending on the end use of the dried timber and the desired quality of the dried timber. Some drying schedules employ higher temperatures and lower humidity, while other schedules call for lower temperatures and medium humidity conditions. Thus, while the use of a certain drying schedule will result in shorter drying times, the use of another schedule for the same species and thickness of timber would result in a longer drying time. The difference in the dried timber, as was pointed out, will be in the quality. It can be stated generally that, based on a certain species and thickness of timber and a desired quality of the dried timber, there is one drying schedule which will result in an optimum drying time. In other words, provided proper kiln equipment and control systems are available, the minimum drying time for timber cannot be reduced without causing a change in the quality of the final product. From this we can conclude that the provision of an optimum amount of heat, humidity, and air flow and proper control systems are vital to obtain short drying times.

Equipment design

This brings us to the subject of kiln design, and basically it can be stated that kilns which are not suitable are those that fail to provide optimum drying conditions or lack in durability or are unable to provide uniform drying conditions over the length and breadth of the load being dried. Most kiln manufacturers today design their equipment in a manner so that the basic heat, humidity, and air flow requirements are obtainable inside the chambers. However, in order not to be superfluous, the design is usually very critical and is unable to provide the basic requirements if the steam pressure falls below the specifications or if there is a drop in electrical voltage or failure of an electrical motor. The decisive criterion, therefore, in selecting a kiln would be the durability of its heat supply system, that is, the fin steam pipe and the air supply units or the fans and motors.

Air circulation systems

Timber, during drying, emits various acids such as tannic acid, formic acid, acetic acid, etc. These acids, mixed with the air flow in the kiln chambers, attack the fans and other equipment inside. Research and experience has shown that iron sheets, even painted zinc sheets or galvanized iron sheets, are most unsuitable for fabrication of kiln equipment. This is because all these metals are very quickly corroded by the acid-laden air and result in defects which necessitate replacements or cause loss in efficiency. For instance, a fan blade made of steel, if corroded at critical points, would cause eccentric operation which would finally result in the breaking away of the blades. It is, therefore, extremely important that the fan inside the kiln is made of 100% aluminium. Aluminium which is 100% pure is completely resistant to wood acids and therefore can last for ages without any defects.

In regard to the position of electrical motors, it is preferable to keep the fan motors outside the kiln chambers. Electrical motors, if placed inside the chambers, are subjected to the hot, humid, and acid-laden air and will suffer frequent breakdowns. Another item inside the kiln which is most susceptible to corrosion is the fin heating coils. This, unfortunately, is also the part of the kiln which one cannot afford to leave in a corroded state or with incrustations and depositions of sawdust on it. The fin pipes are supposed to radiate heat and corrosion or deposition of sawdust on them has the effect of insulation which prevents proper radiation of heat, resulting in a loss of efficiency. Mostly kiln manufacturers use a mild steel steam pipe with an iron strip wound round in helical fashion and spot-welded at some points. The fins themselves are usually corroded very fast, reducing efficiency in the first instance and then, due to constant expansion and contraction of the steel pipe, they lose contact with the main steam pipe surface which results in their becoming absolutely inoperable. Next come the steam pipes themselves, which also corrode speedily, and when there are leaks, the only solution is to replace them. Replacements naturally, besides the expenses involved, cause heavy losses in stopped work or cancelled orders due to unexpected close-downs. Another method is based on crimping an aluminium fin strip on top of a steel pipe. Unfortunately, this system also has not proved very successful because of the fact that moisture and air cause corrosion at any point where steel and aluminium meet each other. This corrosion takes place all along the circumferential length of the joint between the aluminium fins and the steel pipes, resulting in loss of efficiency and, finally, in complete corrosion which makes replacements necessary.

The answer to the above problem is bimetallic finned heating coils made from 100% aluminium and high strength steel. These coils are, in fact, two pipes of different diameters. The steel pipe slides in the aluminium pipe and they are fused together. The steel pipe can stand high steam pressure of up to 20 kilograms per square centimeter. The thick wall of the outer aluminium pipe is grooved by machining to produce a closely placed fin system. The larger number of fins provides the pipe with extremely big surfaces for radiating the heat. The moisture and air does not come in contact with the joint of the aluminium and steel, because the groove is not right down to the steel pipe and a layer of aluminium continues to remain on top of the steel steam pipe; that is, the outer aluminium pipe and fin system which is made of 100% aluminium is completely acid proof. There is no corrosion and no settlement of sawdust on the fins and, therefore, heat continues to be radiated by this system to full efficiency year after year.

Southeast Asian sawmillers are turning to kiln-drying operations to obtain premium prices for their lumber.

If properly stacked and allowed to dry naturally, lumber loses much moisture content before entering the kiln.

CHAPTER 21

Briquetting Waste for Steam and Sale

FRED HAUSMANN
President, Fred Hausmann, Ltd.
Basel, Switzerland

What has John D. Rockefeller to do with the briquetting of wood waste? Quite a lot! The idea of briquetting wood waste is not new. Briquetting had been done—with more or less success—in the 1920s. The idea was derived from the wish to make use of all waste, rather than throwing it away. And it is here that we meet John D. Rockefeller. In his planning and calculations he economized to the last possible cent.

He made money out of waste—a large amount of money. In production of petroleum, Rockefeller made everything into money. What his competitors were throwing away as useless waste, he used. For instance, residues of the distillation process, which to his competitors seemed useless, Rockefeller converted into lubricating oil. Kerosene, another by-product, was sold to light the lamps in China.

Waste has always existed in our society of plenty—always and everywhere. We have ignored it, thrown it away, or burned it uselessly. If John D. Rockefeller were alive today and in my place, he would tell you how much oil he saves daily by making use of wood waste.

As I mentioned previously, the idea of briquetting wood waste is not unique. The following appeared in a 1923 publication:

> The only installation in the United States today for the briquetting of wood shavings for fuel purposes is in Los Angeles, California. The Pacific Coal & Wood Co. is placing the fuel upon the market. The wood waste is briquetted in a special brick press. The pressure obtained is comparatively light. The briquettes are cylindrical in shape, about 3 inches in diameter and 10 inches high, and are finally bound with baling wire to insure their retaining coherence under shipping strains. Though not subjected to extreme pressures, this briquette is reported to be an excellent fuel, the only criticism being that the wire in the ash causes some difficulty in ash disposal.

In Germany, Austria, and even in Switzerland, installations for briquetting wood waste existed more than 50 years ago. But both in the United States and in Europe, these installations had little success and were shut down.

Necessity fostered utilization

Fortunately, in times of war mankind turns resourceful. During World War I, application was made for a patent for a "press for manufacturing logs from damp paper, etc." Figure 21.1 shows the Official Swiss Patent specification and a drawing of the patented press.

During World War II, the idea of briquetting was taken up again in Switzerland in a rather rudimentary way. Between 1939 and 1945, nearly every household owned its little briquetting press, which consisted of a container, a pressure plate, and a lengthened lever arm (see Figure 21.2). Newsprint, soaked in water, was used as raw material. The container had small holes at the bottom and sides, allowing the water to be squeezed out. The log produced had an oblong shape. Briquetting, as a rule, was done by the lady of the house. Drying was left to the sun. We used such a press at home. Figure 21.3 shows pressing dies patented in 1942.

The press shown in Figure 21.4 was designed in Switzerland in 1942. Patents were taken out in Switzerland and England. This machine was the forerunner of today's mechanical extruder.

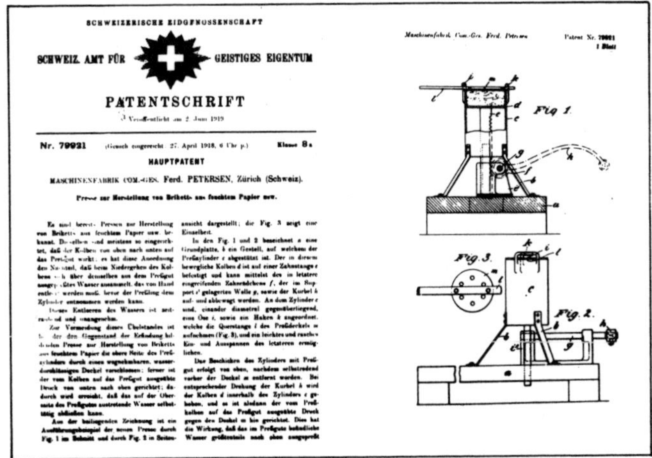

Figure 21.1. Patent specification and drawing of patented press developed for making logs from damp paper. This press was used during World War I.

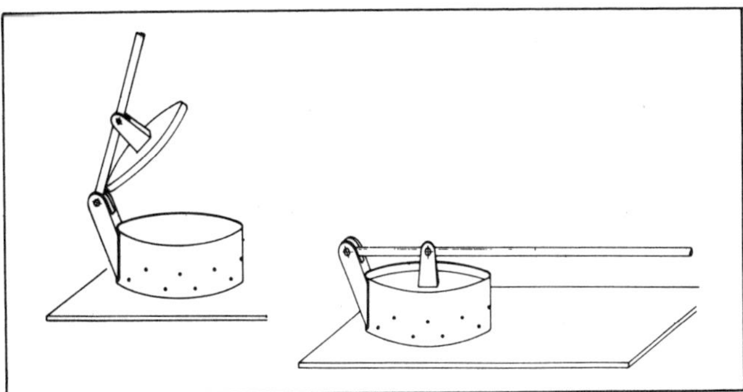

Figure 21.2. A simple household waste paper press used in Europe during World War II. Raw material for the press was wet newsprint.

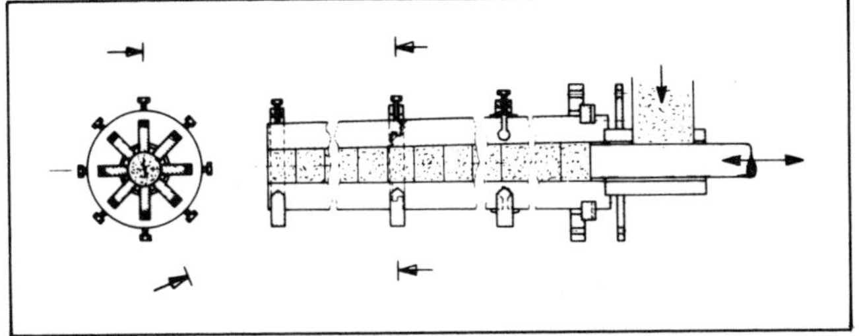

Figure 21.3. Pressing dies patented in 1942.

Figure 21.4. Swiss-designed cellulose waste press, 1942.

The difference between the presses of the 1920s, which in part still survive, and today's Fred Hausmann briquettors is significant in terms of operational sequence. From a technical point of view, particularly regarding the design, progress has been remarkable.

Significant increase in pressing power

For example, the old presses worked with a pressure of about 6,500 pounds per square inch (460 kilograms per square centimeter), while today we work with an average pressure of 17,100 pounds per square inch. On our most recent press of the FH 2-90-200 type, this means a pressing power of 152 tons on the two rams—which is more than 10 times the weight of the machine itself.

Such high pressure requires the very best material. For instance, bearing parts, although not visible, are extremely important. A technically-devised, thorough lubricating system, together with efficient oil filters, is imperative. But 152 tons of pressing power, above all, require the highest precision in construction of the machine. Each of our presses is manufactured with the tolerances valid for machine tool construction. Our system is fully automatic, except for packing.

Briquettes solve European industrial fuel problems

A number of briquetting installations in Europe work without any attendants. Several firms store the wood waste briquetted during the summer months for use in winter to cover their energy requirements. Let me quote an example described in Great Britain's *Timber Trades Journal*:

> No-Nail Boxes, Ltd., subsidiary of Keizer Venesta, Ltd., Britain's largest importer and distributor of plywood, creates in its production processes tons of wood waste in the form of sawdust, small wood blocks, plywood off-cuts, and trimmings. At first, the waste was burned in an incinerator. This not only contravened the Clean Air Act, but as Sir Gerald Nabarro, Chairman and Managing Director, said: "It is almost sacrilegious to waste the heat properties of refined wood products including all the resinous content and vast potential Btu's."

The Hausmann briquettor was installed in the No-Nail boilerhouse as part of a trial scheme to heat the factory from its wood waste. Low-pressure steam from the boilers heats both factories—a total floor area of 180,000 square feet (about 18,000 square meters) at an even 65 degrees Fahrenheit through the winter. Summer output of briquettes is stored for winter use.

Financial savings realized

The financial savings are twofold: (1) no other fuel, such as oil or coal, has been purchased since the briquettor was installed; (2) the high cost of manually removing wood waste from the factories and disposing of it elsewhere is recouped.

The plant and equipment have an estimated life of 25 years, with reasonable maintenance. Sir Gerald Nabarro's own estimate is that, after taking as credit current investment allowances, the entire capital cost of the installation is recoverable in three years. After that period, the heating costs of the factory plus cost of waste collection represent an annual saving in fuel and labor of £5,000 ($11,750 approximately). "I have proven this process during the past seven years. I think it highly viable, economic, and a most attractive financial and fuel economy exercise," says Sir Gerald.

Another example is the Joka Furniture Manufacturing Co. of Austria. By burning its briquettes, Joka accomplishes two things:

1. It is a "self-heating" factory.
2. It supplies the energy for the hot water heating system that kiln-dries the lumber used in manufacturing its products.

It is rare to get the best of both worlds, but Joka appears to be doing just that with its system that heats the plant and simultaneously disposes of all wood waste.

The Caran d'Ache Pencil Co., Geneva, Switzerland, one of Europe's largest manufacturers of pencils, has been recycling wood shavings, compressed in the Hausmann briquettor, directly into the plant's furnaces. Here again the process is fully automatic and plant management expresses complete satisfaction, especially with the low-maintenance requirements.

Using the following illustrations, I would like to explain how a Fred Hausmann briquetting press works.

Briquetting press feeding device

Figure 21.5 shows the feeding device of the briquetting press; it feeds material to the ram and partially compacts it. At the same time, it prevents backward expansion of the raw material during the forward stroke of the ram. This feeding device improves the rate of admission of the machine and with it, its output.

Principal elements of the briquetting press feeding device are:

1. The housing.
2. The auger.
3. The auger drive.

Breakdown of wood elasticity is crucial

In a publication by Albert L. Stillman, which appeared in 1923, it was made evident that briquetting wood waste is

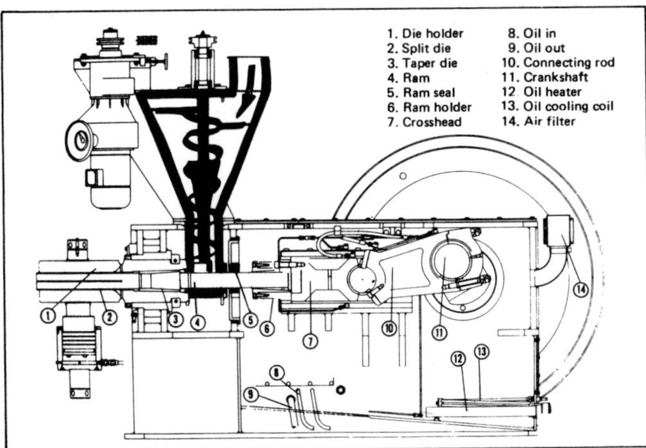

Figure 21.5. Feeding device of briquetting press (shaded area). Shape of the auger prevents the backward expansion of raw material.

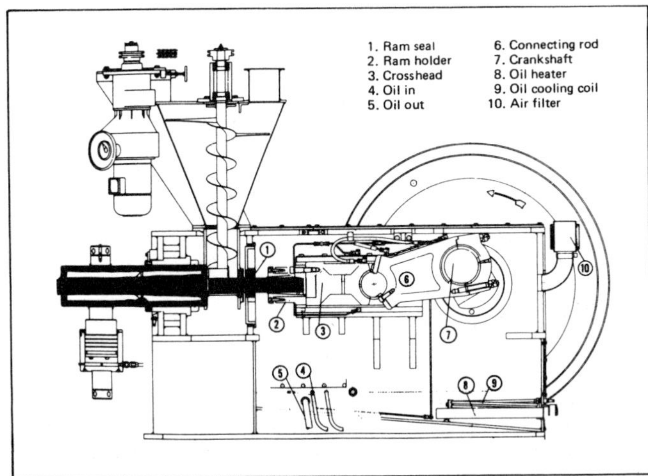

Figure 21.6. Briquettor using wood shavings. Shaded area is die holder, heart of Hausmann briquettor.

possible only if one succeeds in breaking the elasticity of wood. The following is taken from Stillman's report:

> One vital requirement must be met in forming the briquette; namely, the elasticity of the wood must be destroyed either by high pressure or by previous heat treatment, or by a combination of the two, or the briquette formed will have no permanence. Excessive pressures are expensive and can only be justified on something unique and valuable in the final product.

There is no agreement among experts whether the high pressure or the heating up of the raw material is responsible for breaking the elasticity of wood or other cellulose, thus making briquetting, or compressing, possible. Fred Hausmann's presses use both high pressure and heat. The pressure exerted by our ram is 17,100 pounds per square inch. Depending on the raw material and, foremost, depending on intended use of the log, the ideal temperature will vary slightly. Figure 21.6 shows briquettor using wood shavings.

Die holder, the heart of the press

The heart of the briquetting press is the die holder. On the face, it appears to be a thick-walled pipe, but thorough study shows its rich, technical inner life. Here is a short description of the die holder.

1. *Seat and holding device of the interchangeable dies* (locking ring, taper die, split die). As the dies are subject to some wear and tear, they must be changed from time to time for new ones. All our dies are gauge manufactured. They will always fit the machine and easy replacement is guaranteed. The die holder is a patented part of the briquetting press.
2. *Hydraulic power transmission to split die.* The cylindrical split die has four longitudinal slots. By external hydraulic pressure, this cylindrical shape can be changed to a slightly conic form, to increase the friction of the log, thus increasing counterpressure as desired. The four slots of the split die continue outward in the four corresponding slots of the die holder. These slots are conic, to guarantee perfect operation of the hydraulic clamping device. The pressure on the die holder is

adjustable from 5 to 35 tons (11,000 to 77,200 pounds-force).

3. *Die temperature.* During briquetting, the dies get hot. To avoid excessive heat, which would carbonize the log surface, the die holder is provided with a built-in cooling coil. This can be connected to the local water supply. The cooling water remains clean. It can be recirculated or connected to the sewer with no pollution problems.

Figures 21.7, 21.8 and 21.9 show the sequences of operation of three variants of briquetting installations: (1) for fine, dry materials; (2) for coarse, dry materials; (3) for coarse, wet materials.

A simple briquetting installation

The simplest type of briquetting installation consists of a briquetting press and the feeding device. Figure 21.7 shows the feeding of a briquetting press with raw material from a silo through a vibrating chute.

The press also can be fed directly from a cyclone or by any conveying system (such as conveyor belt or screw

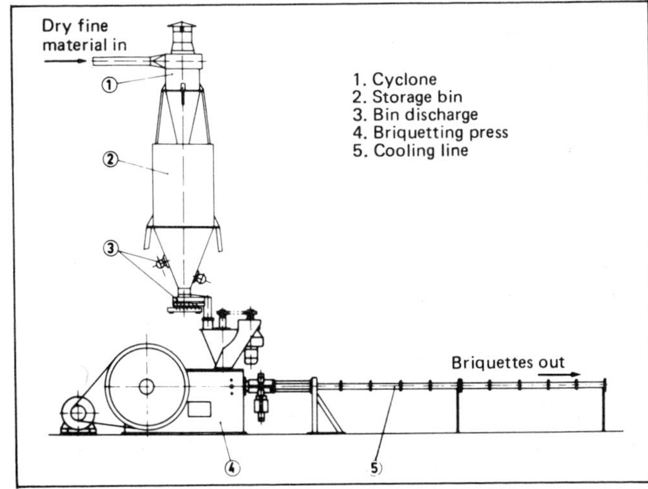

Figure 21.7. Dry fine material is fed from silo through vibrating chute into briquetting press (4). This is the simplest type briquetting installation currently in use.

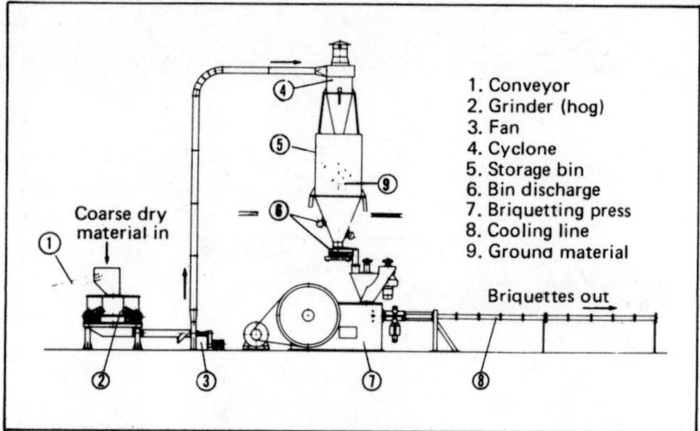

Figure 21.8. A cyclone-to-silo feed for handling coarse, dry material which must be ground prior to conversion into briquettes.

conveyor). In any case, to obtain the best possible results, both in log quality and press output, the feeding of raw material to the briquetting press should be adjustable within certain limits. The type of briquetting installation shown in Figure 21.7 is suitable for briquetting material with a maximum moisture of 15 percent: dry sawdust, sander dust, shavings, peanut hulls or rice husks, etc.

An installation of this type with an output of 2,200 pounds (1,000 kilograms) of logs per hour consumes about 45 kilowatt-hours of power. An installation with an output of 6,600 pounds (3,000 kilograms) of logs per hour, consumes about 105 kilowatt-hours of electrical power during operation.

Briquetting of coarse, dry materials

To briquette dry, but coarse, material, such as trimmings from a plywood or chipboard mill or from a furniture factory, the material must be chipped prior to briquetting. One approach is a grinder, conveying device, bin, bin discharge, and briquettor (Figure 21.8).

Depending on the type of material to be crushed, power consumption of an installation with an output of about 2,200 pounds per hour (1,000 kilograms per hour) is between 85 kilowatts when using a knife-type mill and 120 kilowatts when using a double-rotor hammer mill. Power consumption of this installation, with an output of 6,600 pounds per hour (3,000 kilograms per hour), is between 230 kilowatts and 270 kilowatts—also depending on the type of mill.

Briquetting of coarse, wet material

The operating cycle of a briquetting plant for coarse, wet material is as follows. Wet raw material is hogged and conveyed to a silo. From the silo it is continuously fed into a current of hot air and conveyed to the dryer. A suspension dryer is used in the installation shown in Figure 21.9.

The dried material passes from the dryer to the cyclone and into the briquettor. The hot gas required for drying is produced by burning logs in a furnace. Any conventional furnace built for burning solid fuels can do the job. It is unimportant which type of fire grate the furnace has, or whether it is fed manually or mechanically.

Thermic drying is economically justified only up to 55 percent of the raw material. With higher moisture, as for

instance with wet debarking, we recommend that the material be predried mechanically. Ring roller presses are most suitable for predrying.

Processing engineering data for such an installation (without mechanical drying) with a log output capacity of one ton per hour, are the following:

- Feeding quantity: 3,970 pounds per hour (1,800 kilograms per hour).
- Initial moisture: 50 percent of total weight.
- Heat requirement for evaporation of water: about 4 million Btu per hour (1.0 gram calories per hour).
- Log consumption to produce required heat: about 550 pounds (250 kilograms) per hour of logs.
- Power consumption of installation: between 170 and 210 kilowatts.

The analogous data for an installation with a log output capacity of three tons per hour are:

- Feeding quantity: about 11,900 per hour (5,400 kilograms per hour).
- Initial moisture: 50 percent water.
- Heat requirement for evaporation of water: about 12 million Btu per hour (3 gram calories per hour).
- Log consumption to produce the required heat: about 1,650 pounds (750 kilograms) logs per hour.
- Power consumption: 420 to 480 kilowatts.

Firing and drying

The following explains the firing and drying installations. A furnace for solid fuel has to meet two requirements. First, the fuel must be spread uniformly within the hearth. This can be done by hand (plane-grate type furnace); by chute (sloping-grate type furnace); or mechanically (spreader stoker, traveling-grate, or chain-grate stoker type furnace).

Second, the main combustion air must be distributed in such a way that it is fed uniformly or functionally to the fuel. This is accomplished by installing grate bars or grate plates, air gaps, zones and also by feeding secondary air.

The dryer preferred with our briquetting plant is the

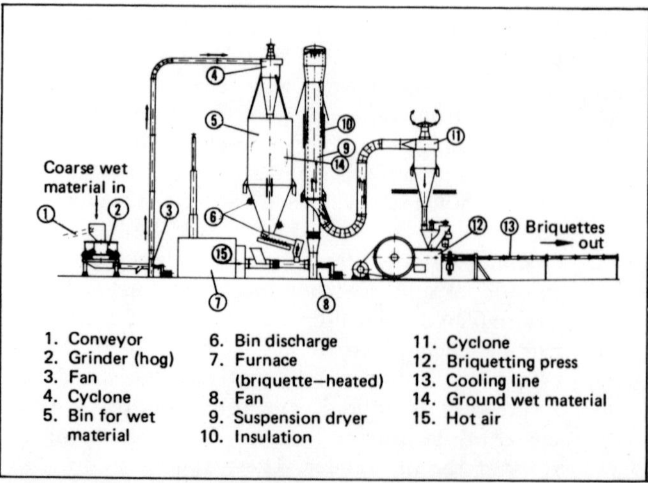

Figure 21.9. Suspension dryer (9) on this installation enables briquetting of wet ground raw materials.

suspension dryer (shown previously in Figure 21.9). It is characterized by simplicity: aside from a fan and an air lock, the dryer has no moving parts.

Suspension drying involves introducing small particles into a hot gas stream. The particles are conveyed and, at the same time, dried by direct contact with the gas. Gas temperatures are normally in the range of 500 to 1,500 degrees Fahrenheit. Contact time between the particles and the gas is short—a matter of seconds. Suspension drying is often called flash-drying because of this short contact time between particles and gas.

Successful drying of sawdust

It might seem that a heat-sensitive material, such as sawdust, could not be exposed to temperatures as high as 1,200 degrees Fahrenheit or more. However, upon entering the gas stream, the sawdust particle contains moisture and the particle tends to stay at the adiabatic saturation temperature of the gas, which is below 212 degrees Fahrenheit, until the moisture is evaporated. Heat for evaporating moisture must come from the gas, thus lowering the gas temperature. By the time moisture is removed from the particles, the gas is at such a low temperature that there is little tendency to char or burn the wood. Of course, particle size is important in performance of a suspension dryer.

In conclusion, let us return to John D. Rockefeller and consider what an enormous contribution we all can make to the solution of the energy crisis. I have desperately tried to get statistical information on cellulosic waste produced in Southeast Asia. Lacking such figures, let me quote some statistics which apply to the United States.

In the four Pacific states of Alaska, Washington, Oregon and California, 21 million tons (dry weight) of unused bark and other wood waste are produced each year. (My source is the November 1973 issue of *Forest Industries*). Twenty-one million tons of briquetted logs equal 81 U.S. gallons of fuel oil per second. Figure 21.10 shows the weight and heat value relationship of fuel oil and logs. Figure 21.11 compares the weight of fuel oil to briquetted logs in weight input and heat value (Btu's). Conversion of that 21 million tons of unused wood waste into electric energy (Figure 21.12) would provide 11 million kilowatts.

The world reserves of raw and other materials are limited. The capability of our earth to absorb waste and pollution without danger is also limited. All over the world, countries try to increase their national production, thus increasing their standards of living. Thus, consumption of raw materials increases, again increasing waste.

Our civilization is based on the exploitation of natural sources of energy. Accordingly, a sure energy supply is one of the fundamental conditions of our economic life. Energy increases our comfort and the material well-being of mankind. We cannot produce more than the earth can give, and the environment can suffer, without danger.

That means: Don't dump it, squeeze it.

---

21 million tons (dry weight) of unused bark and other wood residues are produced each year in the four Pacific states of Alaska, Washington, Oregon, and California (Source: *Forest Industries*, November 1973).

Compressed into 21 million tons of Aglo logs, these 21 million tons of wood waste equal the following amount of fuel oil:

$$\frac{21,000,000}{365 \times 24} = 2,397 \text{ tons per hour}$$

$$= 4,794,000 \text{ pounds per hour}$$

$$\frac{4,794,000}{36,000} = 1,332 \text{ pounds of Aglo logs per second}$$

7,830 Btu/lb = lower calorific value of Aglo logs made from average wood waste

17,640 Btu/lb = lower calorific value of average fuel oil

$$\frac{17,640}{7,830} = 2.25 \text{ pounds of Aglo logs equal 1 pound of average fuel oil}$$

$$\frac{1,332}{2.25} = 592 \text{ pounds of average fuel oil per second}$$

7.344 lbs/U.S. gal = specific weight of average fuel oil

$$\frac{592}{7.344} = 81 \text{ U.S. gallons of fuel oil per second are wasted now}$$

Figure 21.10. Comparison of wood waste recoverable as briquettes with fuel oil consumption in the United States.

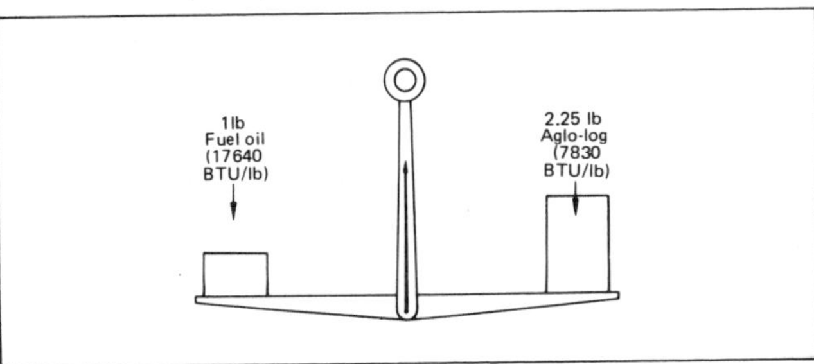

Figure 21.11. Relation of fuel oil and briquetted Aglo logs in terms of weight and heat.

---

21 million tons (dry weight) of unused bark and other wood residues are produced each year in the four Pacific states of Alaska, Washington, Oregon, and California (Source: *Forest Industries*, November 1973).

Compressed into 21 million tons of Aglo logs, these 21 million tons of wood waste will produce the following amount of energy:

$$\frac{21,000,000}{365 \times 24} = 2,397 \text{ tons per hour}$$

$$= 4,794,000 \text{ lbs/hr}$$

Lower calorific value of Aglo logs made from average wood waste = 7,830 Btu/hr

4,794,000 lbs of Aglo logs per hour = 37,537,000,000 Btu/hr
= 37.5 billion Btu/hr

3,413 Btu/hr = 1 kW
37.5 billion Btu/hr = 11 million kW

Generating capacity of the Bonneville Dam (Columbia River) = 564,000 kW

11 million kW = 19.5 times this capacity

Figure 21.12. Potential energy supplied by briquetted wood waste compared with power of Bonneville Dam.

CHAPTER 22

Veneering and Layup of Tropical Hardwoods

TIM A. BAILEY
General Manager, Jurong Plywood Co. (Pte.) Ltd.
Singapore

My topic is very broad, and difficult to cover completely in the space allotted. It is, in fact, the problem to which I have addressed most of my efforts for the past five years, and I guarantee no one will ever have all of the answers. I don't pretend to.

Rather than relate a number of solutions, I would like to approach the subject from the opposite side and relate to you the questions and problems which one addresses when he, in fact, decides to embark on the project of peeling and laying up tropical hardwoods.

The project can be approached in two ways.

1. I have some raw material (a concession) and would like to convert it to a manufactured product.
2. I know of a market for a specific product that I plan to produce.

Project—complete plywood mill

Manufacturing people would naturally be concerned with approaching the project in relation to the first manner I mentioned. That is, we have a concession and we want to convert the logs into veneer and/or plywood.

We would at this point ask a basic question: do we want to make veneer only or do we want to go the entire route and construct a complete plywood mill? This question relates to market studies, shipping availability studies, and capital available. Let's assume we have decided to build a plywood mill.

Equipment

Equipment is the bulk of your investment and requires a good deal of time and a lot of questions prior to purchasing. There are many, many different brands of equipment on the market place today; I assure you that any large equipment company would be more than happy to plan a plywood mill for you. I do, in fact, know of a number of mills in Southeast Asia which were *planned by equipment companies.* I would warn you, however, that unless you are familiar with the production of plywood and the logs that are available, you should be very hesitant of allowing any one equipment manufacturer to build your mill to the specification they feel you need. If you don't know plywood and the raw materials, hire someone who does before you discuss equipment.

The first question before you decide to build your mill or buy equipment should be: how much wood is available for the production of plywood? This may sound silly, but it is, in fact, many times overlooked in the planning of a plywood mill only to find that when completed the mill is:

1. Too large for the wood source available.
2. Improperly designed for the log sizes available.

3. Improperly equipped for the species to be handled.

I assure you that I can take you to at least a dozen mills in the Singapore and Malaysia area alone where there is surplus equipment which has never been used or has been used and found totally unsatisfactory.

Wood source

Know your wood source. Know the *average log size.* This is important. You need to know whether your mill should handle 90 logs a day or 250 logs a day to get the needed output. Know your *species.* Are they readily marketable? Some of the species have characteristics which don't lend themselves to peeling. Other species are not bondable with exterior resins. Can the species be barked mechanically or are they more suited to hand barking? These are the types of questions to answer before you begin to discuss your equipment and design requirements for your plywood mill.

High grade vs. salvage

Another basic question about your plywood mill related to your concession which you should analyze very carefully to determine your optimum return is: do I want to utilize all of the large high grade species on my concession and make the highest grade panel possible, or do I want to export all of the very high grade desirable logs possible and utilize my veneer mill or plywood mill on the *non-exportable logs?* Either system can be profitable.

Wood species

Species of wood to be used within your mill are very important when selecting equipment. Some tropical species are soft and easily workable, whereas others are so hard that they must be steamed prior to peeling. Certainly the great majority of standing timber in Southeast Asia is of the *meranti* group. However, there are a great number of other species which have very desirable properties for plywood production if one is aware of how to handle them. Ideally you should have all of this information before you begin to shop for equipment. If you don't, try to determine which equipment offers the greatest possible flexibility from your standpoint.

Here are four very general questions to ask yourself about the equipment:

1. Is the equipment capable of handling very heavy logs adequately as well as normal size and weight logs or blocks?
2. What is the maximum output? Does this piece of

equipment have the flexibility to run at high speeds as well as low speeds?

3. How many automatic devices, electrical devices, etc., are there on the various pieces of equipment which could cause future maintenance problems?
4. What is the manufacturer's parts supply policy and his service policy?

The above four questions, if you carefully analyze and answer them correctly, might point out that, strangely enough, the *cheapest equipment* available may not be the *least expensive.* Conversely, the supplier stressing how heavy, how fast, how automatic his equipment is as being his reason for his higher price may not be the profitable choice, either. For example, some particular piece of equipment is quite capable of handling logs and blocks weighing 12 tons. If the great majority of your logs only weigh 4 tons it wouldn't make a lot of sense to install such heavy equipment at a higher cost.

Equipment output and balance

Another very important series of questions to ask yourself relate to the output balance of the equipment being offered. For example, three 20-opening hot presses won't do you much good if your lathes can only produce sufficient veneers to keep two of them busy. Problems such as this are quite common in hardwood mills in Southeast Asia. These questions should be approached from the very outset of your project. With planning and forward thinking, future balancing or expansion can be quite painless and profitable, whereas without serious thinking, it can later be quite painful and quite costly.

Panel mix

Another very important item to keep in mind in considering the balance of the plywood mill is panel mix. You can't simply ask someone to design a mill to produce 30,000 cubic meters of plywood annually. I am sure it will take considerably different equipment to produce 30,000 cubic meters of 18mm plywood versus 30,000 cubic meters of doorskins. You must think in terms of pieces to handle as well as volume. 30,000 cubic meters of 18mm plywood is a little over 500,000 panels, whereas it is over *3.5 million doorskins.*

Construction

Once your mill is planned in general terms and the equipment purchase is decided upon, the construction phase begins. The right questions here can pay off also. I would like to list a few:

1. When do I need the equipment, considering installation time, etc?

2. Should I go all out to start the lathes and sell veneer to get some cash flow going or if the delivery times don't fit, should we start our press first and purchase veneer for a while?
3. Have we planned adequately for waste removal?
4. Have we designed the steam, electric and air system with sufficient capacity for some expansion?
5. Should we hire some of the key people right away so they become a part of the planning process and gain familiarity?
6. How can we arrange the mill to get the *best supervision* possible?

Plant shape

Mills in Asia are traditionally constructed in one straight line with the result that the finish end can be 1/4 to 1/2 mile from the green end whereas a very few mills have chosen a "U" shape with the distance between the various departments greatly reduced.

Production

The day will come when you start your mill, and the question which will continue to haunt you for the next six months to a year will be: why did I ever decide to go ahead with this project?

If you asked the right questions from the outset, and if you come close to answering 60% of them right, sometime within a year after your start-up, your mill will begin to show a profit and you will begin to grasp the whole business more and more each day. You will constantly have new ideas and your question from then on will be: why didn't I think of this before? If you didn't answer at least 50% of the initial questions properly, your mill won't improve and eventually you will ask: where can I unload this operation?

I have tried to mention some basic questions regarding the construction of a mill and selection of equipment. I have not offered any answers, but I can suggest where you get the answers. The easiest way to get your answers is to let someone else do it. Technical advice can be purchased.

1. You can hire someone who has the experience you need to plan, build, and run your mill.
2. You can joint venture—you put up the capital, they put up the expertise, and you both share in the profits.
3. You can spend what time you feel is necessary visiting mills asking questions and then attempt it on your own.
4. Find a successful operation and copy it.

I might mention that the last two alternatives have proven the least successful.

CHAPTER 23

Training Programs for Management and Operating Personnel

LEONARD H. TAYLOR
Consulting Engineer
Vancouver, British Columbia, Canada

The purpose of this chapter is to emphasize the importance of adequate and continuing training of all operations workers and of members of the sawmill staff. Many plants overlook this facet of the complete operating picture, with resulting losses in revenue and staff stability, wastage of supplies and of raw material, injuries to employees, and damage to equipment.

Lethargy of routine operations

It is easy to operate a plant on a day-to-day, humdrum basis and to incur various losses and wastage without realizing that a small investment in training often pays large dividends. A thousand dollars spent on training can often return $10,000 to your year-end revenue. I will give some actual examples of this later on. Each operation, the duties of each employee, both on staff and in the operations section, must receive critical analysis from time to time to see if any improvements can be made. Just because a job has been done in a certain manner for a long time is not reason enough for continuing with it at all or in that manner.

Skill, training, and experience of sawmill employees

Figure 23.1 shows four columns of sawmill employee titles divided according to the amount of training and experience required to do their work. The category of the plant operators is, of course, dependent on the type of mill equipment involved. How do you rate your staff compared to this tabulation? Do you allow enough time for training?

High costs of lack of training

Training pays off in handsome dividends in many places: head sawyer, graders, saw filers, salesmen, and accountants. Here are two simple but realistic examples:

Figure 23.2 shows the high cost to your plant for the lack of training of an edger man. Let us assume that the edger man is not familiar with the wane allowances and grading rules regarding wane in dried and planed boards. Suppose, also, you are cutting mainly 2-inch lumber about 15 feet long and the edger man processes five boards per minute for 400 minutes per shift. Now if he mis-cuts 10 percent of these boards by making a 2x4 out of a piece that should have made a 2x6, or he makes a 2x6 instead of a 2x8, he will throw away 200 pieces of 2x2 lumber per shift. In 250 shifts per year he will waste 250,000 f.b.m. of lumber, which at a nominal figure of $S*600/M f.b.m., will cost you $S150,000 in loss of gross revenue.

How much does it cost you to train that edgerman so that he reduces his errors by 75 percent? What equipment improvements can you make to help him—shadow lines, better feed-rolls, board-flippers that enable him to turn a heavy plank easily to look at the other side? The economics

* Singapore dollars

*OVER 6 MONTHS	1-5 MONTHS	3-4 WEEKS	LESS THAN 1 WEEK
*Mill Manager	Lumber Grader	Gang Sawyer	Lumber Piler
*Production Superintendant	Pony-rig Sawyer	Edgerman	Janitor
*Sales Superintendant	Crane Operator	Fork Truck Operator	Utility Man
*Mill Foreman	Electrician II	Cut-off Sawyer	Clean-up
*Head Filer	Mechanic II	Trim Saw Operator	Guard
Maintenance Supervisor	Filer	Trade Helper	Driver
Head Sawyer	Resaw Operator	Greaser	
Electrician	Log Truck Driver	Filer Helper	
Mechanics	Log Fork Truck Operator	Tail Sawyer	
Welder	Clerk	Lumber Truck Driver	
Shipper	Stenographer	1st Aid Attendant	
*Office Manager	Bookkeeper		
Secretary			
Accountant			
Personnel Manager			
Sales Clerks			

*The time required to train these men will be several years depending on their practical experience, the complexity of the operation, and their own ability.

Figure 23.1. Training and experience necessary for sawmill employees.

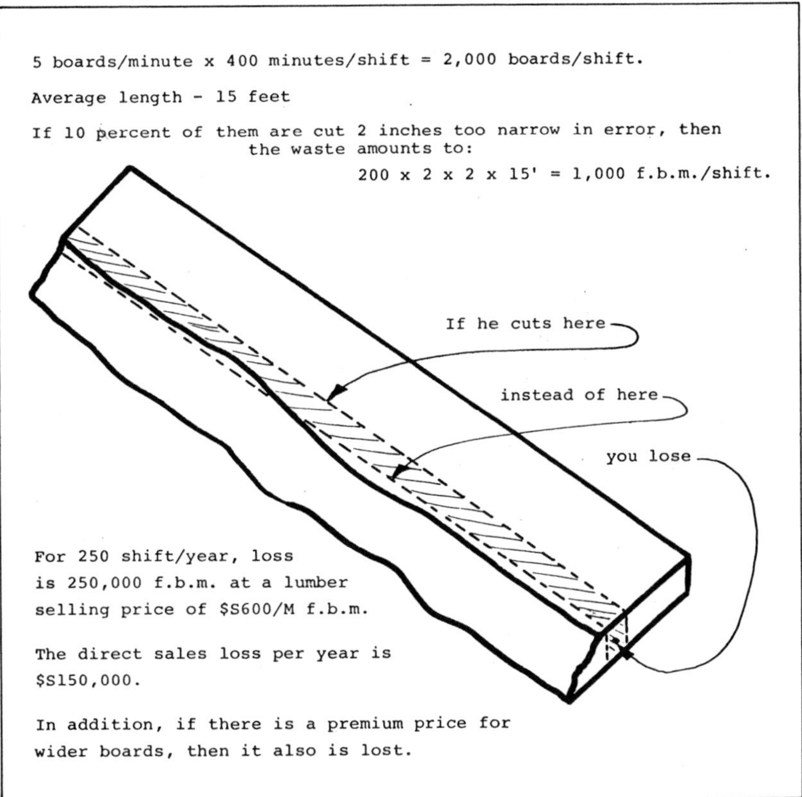

Figure 23.2.   Loss from incorrect edger operation.

of success or failure are very close. Turning that board may be the difference between a profit and a loss to your mill.

If lumber is required to be trimmed to one-foot increments or to two-foot increments, then losses are often incurred at the trim saw in cutting back to the next length because the log was a few inches short of making the next longer length. So log bucking at the millsite can be costing you more money than you'd care to learn about. Cutting logs to minimize the impact of log defects is also important. So, here again, training and good measuring equipment will often pay for itself in a short time.

Figure 23.3 shows a log with no defects which must be cut into three pieces for transport and sawing in the mill. The dimensions shown above the log are ideal for giving a few extra inches for lumber trimming, and should enable the mill to produce 12-foot, 16-foot and 18-foot boards.

Suppose the bucker in the woods, or at the mill landing area, cuts on the lines indicated by the dimensions below the trunk. These are not big errors—only 8-inch on the butt log and 6-inch on the top log. Not really a serious mistake, but look what it does to your log costs if your mill is selling lumber in even-foot increments.

All the boards from the butt log and top log are trimmed back to 16 and 12 feet as they would be for the ideal bucking lengths. However, the lumber from the middle section of the log is trimmed back to 16 feet instead of 18 feet. You have thrown away two feet out of a 47-foot-8-inch log or about four percent. Translate that into a language that really talks—money! If your mill is cutting 200 cubic meters of logs per shift for 250 shifts per year, you are consuming 50,000 cubic meters of logs per year.

If this type of error is made on 10 percent—only 10 percent—of the logs and your recovery of lumber from the logs is 50 percent (a bit optimistic, but easier for the arithmetic) then you are losing 100 cubic meters of lumber (50,000 x 10% x 4% x 50% yield). 100 cubic meters of lumber are equivalent to 42,000 f.b.m., which at the nominal selling price of $S600/M represents a loss of $S25,200. You have to be a wealthy man to be able to afford *not* to give adequate training and supervision to the loggers who are cutting your timber to length.

Recommendations

Such numerical values can be calculated for most operations in the mill provided sufficient data on actual log deliveries and lumber production is available. Most savings are not so spectacular as these, but, nonetheless, making the calculations is worthwhile.

In the case of the edger man, he must have a training session with the grader and make regular examinations of the output of the planer to see the importance of leaving some wane on boards. He must also be advised daily on the main widths of lumber required to fill current orders.

Make sure your loggers have some appreciation of the needs of the mill and that they cut the trees into log lengths that suit the mill operations and order file. Make periodic summaries of log lengths as received at the mill, indicating those that are not within the limits you have set, and discuss the statistics with the logging manager.

Follow-up training

Training has to be continual in any active plant, usually on an informal basis, such as the foreman reviewing the worker's product and procedures during the normal course of the work to make sure he continues to follow his training guidelines. More formal and definite training must

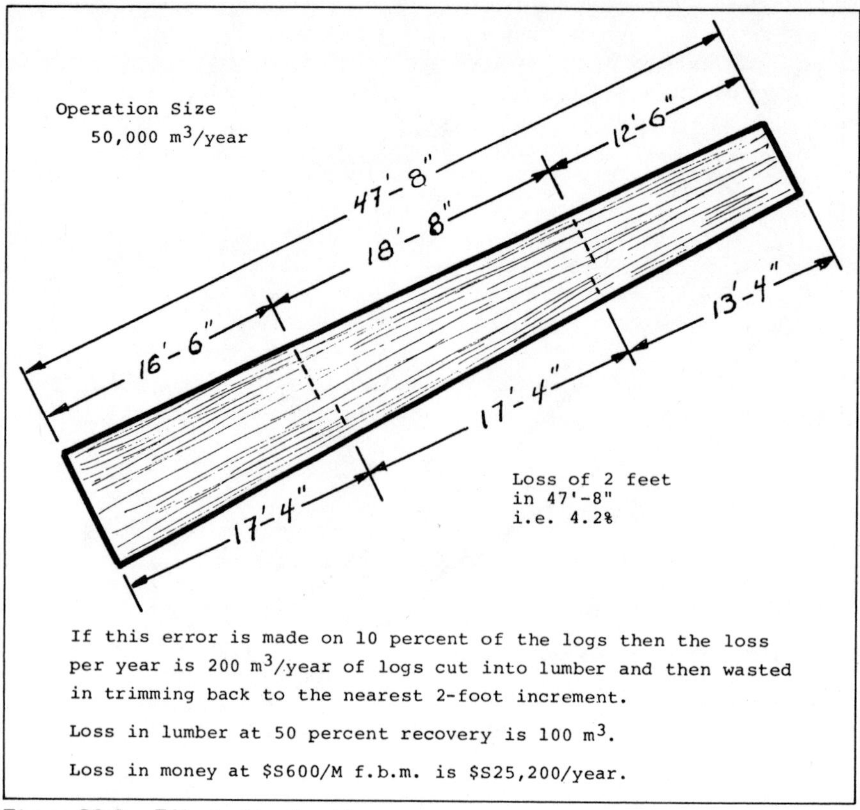

Figure 23.3. Effect of log bucking error.

be given to new employees, to all employees involved with new products, new procedures, or new machinery. Don't take it for granted that a machine operator will soon be familiar with running the new machine just because it is similar to the old one. Small differences suddenly become very significant in cases of emergency. Give him time to look it over; show him all new features, the use of guards, the controls, and the lubrication required; try to have the operators help install the machine and try it out for an hour or so before the start of the production shift.

Equipment manufacturers will teach your crew how to operate and maintain the equipment they install, but not how to make a profit with it. Each operator must be familiar with the lumber he is manufacturing to get the best utilization from his stock; he should not be just a machine operator who pushes the lumber through the saw.

Make a lost time record of each major piece of machinery showing the reason for stopping and the number of minutes for each halt. With this record, trouble areas are clearly shown. For example, if the headsaw is stopped frequently because of lack of logs, then you have to improve the supply system to keep it busy. Each operator can have a report form near him to note time of stopping, reason for stopping, and time of re-starting. Record these figures and keep the operators' interest by discussing the reports with them to emphasize the need to keep the plant operating. When a saw is just cutting air it is not making any money for you.

The lost time reports will also indicate maintenance problems. If the equipment is functioning poorly, then look to your maintenance procedures and maintenance staff training. See that troublesome areas get the attention they need during lunch breaks and between shifts, so there is no lost time during the plant operations. A well-trained and conscientious maintenance crew are a joy to any mill operation. But they must have adequate tools, training, and supervision so they can make running repairs quickly and well, even if only well enough, in some cases, to carry through to the end of the shift.

This lost time procedure is more applicable to the type of sawmill we are accustomed to running in North America and Europe, wherein the whole mill runs as a continuous unit with some flexibility in the storage conveyors between the major machines. In Southeast Asia, the mill machinery is often not designed for continuous cutting, but for treating each log as a separate project.

In many North American operations, the mill manager works out the value of the sawmill in dollars per minute of operation. He may use dollars of gross revenue, or operating cost dollars, or another basis that suits his purpose. But when he has a figure of, for example, $26 per minute of gross revenue for the operating mill, and an average lost time in the mill of 100 minutes per shift, he is going to analyze the source of this daily loss of $2,600. If he doesn't make some improvements by better training and perhaps some equipment improvements, he can expect the president of the firm to find another manager who can reduce these losses.

Here again, I want to emphasize that operator and staff training usually result in large improvements without equipment changes. In other words, review your individual problems but be slow on blaming the equipment. It is usually well-designed and, if operated and maintained properly, will produce well for you.

Office and management

Analysis of the productivity of office and supervisory

personnel is not as directly tied to a production that is easily measured like cubic metres per shift. A well-trained, well-occupied staff is a very important part of the operations. Avoid the fault of hiring too many poorly-trained people instead of fewer who are well-trained. No matter how many clerks you hire, they will always appear busy. Four or five clerks doing only one-half a day's work each, are a waste of your money and office space, and they decrease the efficiency of the rest of the staff. Periodic reviews of the work load and training requirements are important here also. Management or staff errors are made due to lack of experience and training. Some errors are expected, but the average must be good. The old adage applies that if a man doesn't make any errors, then he probably isn't making any decisions either.

New plant design, construction, and operation

A new project will probably include most of the steps shown in Figure 23.4. In most cases, each step is indispensable to the successful completion and satisfactory operation of the project. Training is one of these items.

When new equipment in an old plant or a whole new plant does not come up to production expectations, don't jump to the conclusion that the machinery or layout is completely at fault without having a critical look at your management, your operators, and your maintenance training expertise.

Some operators require a relatively short training period—perhaps only a few minutes on the job—but virtually all operations require continual supervision. Other operations take weeks, months, and years of training to acquire the skill and judgement needed to fill the position adequately. For example, the saw filer, or saw doctor, in North America and Europe, works for at least six years at the trade before he can hold the job as head filer in a medium-sized mill. We consider him to be one of the most important men in the mill operation, the mill manager not excepted. So, if you require a head filer trainer to assist your mill at start-up time and to train your filers, don't expect him to do this job in three weeks. If your mill is cutting a variety of timber species with a high capacity bandmill, you may need to budget for such a training filer on your staff for many months, depending, of course, on the previous experience of your employees. You won't benefit from your new plant without him.

Other tradesmen and operators usually require less time, depending of course, on the complexity of the operation and the experience required. But get *good advice* on this very important part of your operation and include the costs in your *capital budget*.

Following are several of the many methods of giving

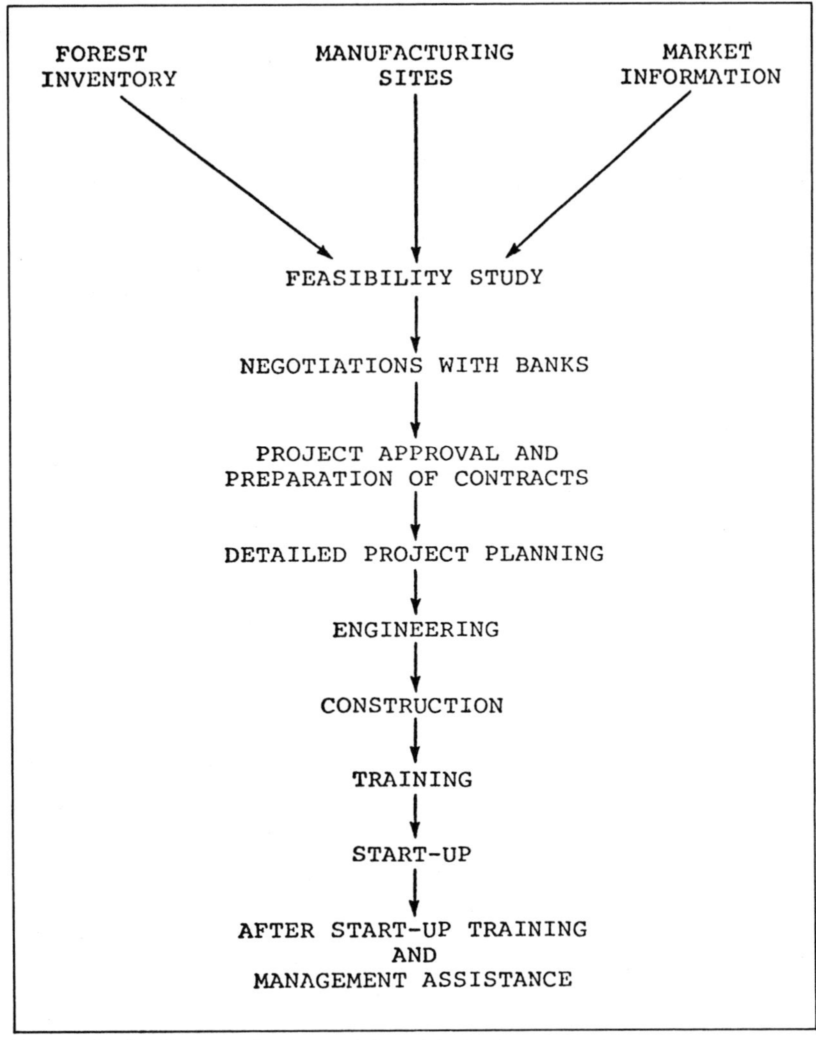

Figure 23.4. Stages of a typical forest industry project.

training and from this list you will choose those methods that suit your conditions. Most likely you will use a combination of these methods and will, of course, vary the combination for the various levels of employees.

1. Classroom talks.
2. Manuals.
3. On-the-job training.
4. Night courses while working.
5. Equipment manufacturer's courses.
6. Work at similar plants.
7. Overseas training.
8. Refresher courses.
9. Any combination of these.

C. D. Schultz and Co. have usually worked on overseas training jobs to give direct training to the local counterpart trainer who will, in turn, train the workers. With this procedure, the cost to the host company or country is much less.

Item 7 above—the overseas training—is only justified for highly specialized personnel and normally should be avoided as being very expensive and difficult to control.

Here again, review the many alternatives and draw up your programme for your particular operation. In planning your training programme, you should:

1. Define scope and participants.
2. Select training supervisor.
3. Select method.
4. Prepare training material.
5. Arrange schedules.

Set up a plan of attack on this very important part of your operation. Like any other phase of your operation, it will pay you more benefits if you plan it carefully. Discuss the needs and objectives with your plant supervisors, and get the reactions from the employees themselves about the utility of the course after they have been exposed to it.

Problems

I want to review briefly some of the problems encountered in many parts of the developing countries in setting up a plant and in training a new crew to operate it.

Poorly qualified and inexperienced groups

Many plants are being established, frequently with the help of governments, to develop industry in poor rural areas that have little to offer the residents now, except a marginal existence. Such plants are not located in that area on a strict economic basis, but to fill the sociological needs of the residents. These efforts to bring some of the benefits of economic development are worthwhile, but the sponsors must realize that there are also operational and training problems involved—i.e.: additional costs, slower realization of production capabilities, difficulty in staffing with experienced supervisors, etc.

Unwillingness of educated people to work

Many, perhaps even the majority, of university-educated men in developing countries are extremely reluctant to get their hands dirty. This factor is a very serious drawback and a problem to the training and operation of a plant. It is a complete reversal of the attitude of many of the professional graduates from North America, and is particularly serious since the North American university student usually obtains practical working experience in between his study semesters, whereas many Asian students do not work. The problem is that the school system does not allow several months in between the academic years for work. This situation should be changed, or the student himself should take a few months off every year or two to get practical experience doing useful, and even manual, work.

Therefore, the North American forester, for example, can and has done virtually every job in the forest operation: survey crew chain man, axe man, surveyor, bulldozer operator, mechanics helper, faller, bucker, scaler, skidder operator, truck driver, time keeper, and operations manager. He can demonstrate and teach every facet of the job, even in the pouring rain, and ankle-deep in mud. He has no illusions about a university degree suddenly elevating him to an ivory office from which he can issue directives and expect the work to go well.

This is one of the most serious problems we, as trainers in the developing nations, have to face, and until the governments and major employers of university graduates change their attitudes and expectations of these graduates, it will continue as a problem.

Interference of owners and government officials

Even when a plant is running well, the management may be instructed to make some changes in the operation to suit some vague plan of government officials whose background in the industry may not really qualify them to make that decision. Basic changes in the operations should be left to the qualified management, and suggestions should certainly be discussed with officials without any dictatorial and uneconomic pressure.

Poorly qualified trainers

A trainer must be well qualified with a wide variety of experience and not just someone capable of reading a service manual. Employees learn by doing—by seeing the job done by an experienced man, by copying him, by asking whys and wherefores and getting logical straightforward answers. Your employees don't get this assistance from an inexperienced trainer no matter how well-meaning his intentions may be.

Lack of commitment

The main purpose of a professional trainer—i.e., a man brought in to give specialized training, as distinguished from a full-time training director employed by the operation—is to work himself out of a job. He must be committed to pushing his local counterpart into taking full responsibility of the work as fast as he can handle it. If you are using training specialists, keep a watch on this point. In the case of equipment operators, your employees should be running the machine most of the time with guidance from the trainer. In management functions all decisions should be made after a preliminary discussion with the local personnel, while the trainer may offer suggestions and recommendations on the subject; however, the trainer should probably have the veto power or overriding authority during the first few months of his contract.

Cultural shock of foreign trainers

Many North Americans or Europeans who come to a completely different region, like Southeast Asia, are overwhelmed by the great differences in the way of life. The climate, the large masses of people, the different languages, the change in religious concepts, the strange food, etc., give some of us a problem in functioning for a few days or weeks even. However, this is primarily a problem of the hiring procedure. All potential overseas employees must be well informed as to the conditions of life in the host country and should read about the recent history, review the customs of the people, become familiar with the currency and travel regulations, and learn a few words of the language. If a man is to bring his family with him on a contract, then his wife and children should also be interviewed and prepared for their new environment, in order to avoid sending a family overseas and having them quit after a few months because the wife cannot cope with the new life.

In one company I worked for, the employee could not bring his family overseas until he had been there himself for six months, during which time he could decide whether he wanted to stay several years, and whether his family would like it. In the meantime, he learned some of the language and customs, and was prepared to receive his family. Also, the company could decide whether or not it wanted to keep him before going to the expense of moving him.

Conclusion

Training is just as much a part of plant operations as concrete foundations and machinery, and should be capitalized in your budget. Give it at least as much attention as you give to equipment selection and plant layout. Don't build a plant and expect it to run by itself.

Exhibitors
Southeast Asia Sawmill Seminar
June 1975, Singapore

The variety of exhibits and the proximity of the exhibit area to the program auditorium were important ingredients in the success of the Singapore Seminar. Clinic registrants had the opportunity to speak at length with the exhibitors' representatives—to discuss particular equipment or machinery and to talk over various problems at length.

It is impossible to include in the proceedings the many useful ideas and specific information exchanged, but there is no question of the value of that information. The exhibitors' support of the Seminar is appreciated. Here are the companies who had booths at Singapore in 1975:

Armstrong Mfg. Co.
Black Clawson, Inc.
Burmah Far East
Caterpillar Far East Ltd.
Coe Mfg. Co.
Columbia Helicopter, Inc.
East Asiatic Co., Ltd.
The Ellison Co.
Filer & Stowell
Folin & Brothers Sdn. Bhd.
Robert Hilderbrand Maschinenbau GmbH
International Enterprises of America, Inc.

Lahden Rautateollisuus Oy (Rau-te)
Leong Bee & Soo Bee (Singapore) Pte. Ltd.
Mark 50 Machinery
Mobil Mfg. Co.
Niven (Asia) Sdn. Bhd. (Pace Products)
Poclain S.A.
Powell Machine Works Ltd.
Sandvik Singapore Pte. Ltd.
Simonds Cutting Tools
Spotradio Sdn. Bhd.
T-Export (Hamburg) GmbH
Vollmer (Singapore) Pte. Ltd.

Yates-American Machine Co.